Japan and World Timber Markets

Japan and World Timber Markets

P.R. Blandon

Department of Business Studies
London Guildhall University
UK

CABI *Publishing*

CABI *Publishing* is a division of CAB *International*

CABI Publishing
CAB International
Wallingford
Oxon OX10 8DE
UK

Tel: +44 (0)1491 832111
Fax: +44 (0)1491 833508
Email: cabi@cabi.org

CABI Publishing
10 E. 40th Street
Suite 3203
New York, NY 10016
USA

Tel: +1 212 481 7018
Fax: +1 212 686 7993
Email: cabi-nao@cabi.org

A catalogue record for this book is available from the British Library, London, UK.

Library of Congress Cataloging-in-Publication Data
Blandon, Peter, 1949-
Japan and world timber markets / P.R. Blandon.
p. cm.
Includes bibliographical references and index.
ISBN 0-85199-327-3 : (alk. paper)
1. Lumber trade--Japan. 2. Lumber trade. I. Title.
HD9766.J32B58 1999
382' .45674'00952--dc21 98-56385
CIP

ISBN 0 85199 327 3

Printed and bound in the UK at the University Press, Cambridge, from copy supplied by the author.

Contents

Acknowledgements

The work in this book really began when Professor Nagumo of the University of Tokyo was kind enough to allow me to join his research laboratory in 1983. It was there that I was introduced to the forecasts and forecasting methods employed by the Japanese. Whilst at the University of Tokyo I made the acquaintance of Dr Tanaka, now at Mie University, who helped in my work in forecasting and on whose ideas a large part of the forecasting effort in Chapter nine is actually based.

During my time at the University of Tokyo I often travelled to the Forest and Forest Products Research Institute at Tsukuba where I had a number of useful meetings and conversations with the experts working there and especially Dr Kumazaki. Later I was to work with Dr Kumazaki during two periods of teaching forestry economics at the University of Tsukuba. At Tsukuba I was based in the department headed by Professor Akaha who afforded me every opportunity to further my research. Also, the other members of staff there, especially Dr Narita and Dr Kato, patiently explained a number of ideas and concepts, as well as giving me a broader background in Japanese forestry.

Some of the work here was contained in my doctoral thesis which Dr Colin Price of the University of Wales had the dubious honour of supervising. More recently, Professor Okada of Iwate University has been very helpful in looking at parts of the early drafts of this book and correcting errors, and the officials at the Ministry of Agriculture in Japan were also kind enough to let me take up their time in discussions of various ideas.

Not least, thanks are due to my wife Mariko who has helped to explain some of the papers referred to in this text. The usual disclaimers, of course, apply. Whilst a large number of people have been kind enough to help me in the course of my work, the errors of fact or interpretation contained herein are mine.

Japanese Names and Conversion Factors

In this book, Japanese names are transliterated using the Hepburn system. Long vowels are shown by duplication of the vowel or by *ou* in the case of a long 'o'. The macron or circumflex, which are other commonly used methods, are not employed here. However, where words have an accepted English spelling this spelling will be used. Thus, for example, Tokyo is written in the normal way rather than the technically correct transliteration of *Toukyou*. All transliterations are italicized except for ones that are in standard English use or where they are very frequently used, such as the timber names 'sugi' and 'hinoki'. Here repeated italicization would probably prove irritating. A glossary of the Japanese terms used is provided at the end of this book, starting at page 225.

There are two main sources of data used in this book, FAO and Japanese government sources. The former is sufficiently well-known to require no explanation. However, the Japanese government tends to issue two basic types of data sets. The first, usually emanating from the Treasury, refers to external trade and is quoted in actual units whereas the second type of data, originating from the Ministry of Agriculture, Forestry and Fisheries, often quotes volumes in roundwood equivalents (RWE). The conversion factors used vary slightly from year to year but the latest set, from the Forestry Agency's *Ringyou Toukei Youran* for 1998 are given below.

Item	Factor	Item	Factor
Coniferous sawnwood	72.7%	Pulp	1 tonne = 3.3 m^3
Broad-leaved sawnwood	62.1%	Hardwood chips	1 tonne = 2.219 m^3
Plywood	67.1%	Coniferous chips	1 tonne = 1.835 m^3
Blockboard	40.9%		

Introduction

In books on the forestry sector written by the Japanese it is not uncommon to see some assertion to the effect that Japan has a society based on wood and forests. To quote one such, an introduction from a publication of the Japanese Forestry Agency, '.. the Japanese have established a unique living-style that is referred to often as the *culture of wood*'.

However, the casual visitor to Japan could be forgiven for finding such an assertion somewhat surprising. While the more traditional areas and tourist attractions are indeed characterized by wooden houses, this is not very much different from the wooden buildings found in the old quarters of historic European cities. Similarly, a train trip from, say, Tokyo to Kyoto does not give the impression that around two-thirds of the country is forested. Indeed, a trip of similar duration from London would probably give the overwhelming impression that Britain is more heavily wooded, when in fact Britain's forest cover is only 10%.

On the other hand, import figures certainly suggest that Japan does, indeed, have a heavy dependence on wood and timber products. According to FAO figures for the mid-1990s, Japan accounts for a little over a third of all of the imports of industrial round wood in the world, about 40% of all tropical log imports and nearly two-thirds of all imports of pulpwood and particles.

Being a large importer of a commodity is not necessarily noteworthy unless, of course, the sources of those imports are uncertain and here Japan is probably facing increasing difficulties. A large proportion of its coniferous imports comes from the USA and Canada where environmental concerns are putting pressure on timber harvests. Some of Japan's softwood log imports originate in Russia and here the social and political upheavals have meant that the output of forest products in Eastern Siberia and the Russian Far East has dropped dramatically in the last decade. The volume of tropical hardwood log imports from Southeast Asia has been diminishing for some time as the resource base contracts and the countries concerned move to the export of more value-added products.

However, the importance of Japan in the world timber trade tends to conceal the fact that it is at the same time a very heavily forested country. As already mentioned, its forest cover is about 67%, a figure only slightly lower than those of Finland and Sweden and significantly greater than those of Canada, the USA and Russia. This level of forest cover is all the more surprising when the high population density of Japan is considered. Japan is virtually alone in the industrialized world in combining a high forest cover and high population density. The combination of high forest cover and high imports, especially of tropical hardwoods, means that Japan is often criticized for not felling its own forests whilst causing other countries' forests to be cleared.

The reason for this situation is not that Japan's forests are worthless; they are made up very largely of valuable coniferous species. Neither are they too young to enter the market, a large part is approaching maturity. In fact there is a burgeoning potential supply which is the result of the replanting that was undertaken after the war. This replanting was needed to restock the forests devastated by overcutting in the war effort and, in the immediate post-war years, to restock areas felled for fuel and for housing reconstruction.

A 'back-of-the-envelope' calculation gives an idea of the scale of Japan's maturing forests. The area of man-made forests aged between 46 and 55 years is set to increase about sevenfold over the next 20 years. If the coniferous species that make up almost all of these man-made forests were felled at their usually quoted economic maturity there would be a similar sevenfold increase in domestic timber supply. This would be more than enough to satisfy Japan's present domestic demand and the entire coniferous log imports of the rest of the world, converting Japan from the world's largest importer of timber to its major exporter.

However, it would be a brave person who would forecast such events. The Japanese are more modest in their prognoses, forecasting that Japan might enter a *kokusanzaijidai*, or era of self-sufficiency, when the self-sufficiency ratio will increase from its present level of 20.5% to around 31%. But domestic timber output has fallen by an average of about 2% each year for the past two decades and, when this is considered, even the modest target of a 31% self-sufficiency ratio begins to look optimistic. Part of the reason for the reduction in domestic output and the increasing dependence on imports lies in the nature of the rural economy and part is to be found in the nature of the forests themselves.

This book examines these problems. The question posed is what will be the level of Japanese domestic production in the next twenty or so years and, therefore, what will the level of imports be? To throw light on these questions, if not to provide answers, an econometric model is developed and presented. However, this book will not be an overtly statistical one, only in the final chapters, when the factors impinging upon Japanese and world timber markets have been discussed, will the econometric model be introduced. It will be seen that the results of the model are very sensitive to changes in the data referring to those areas of the Japanese market where, unfortunately, data are at their most unreliable.

The structure of the book is, then, fairly straightforward. Chapter one will look at the background of Japanese forestry and the history of Japan's increasing dependence on international trade for its forest products. Chapter two will look at the patterns of demand for the various forest products in Japan and the factors that influence them. Chapters three, four, five and six examine the supply side of the market. Chapter three continues the Japanese theme by looking at domestic supply. Chapter four looks at the overall position of Japan in the world timber markets and Chapters five and six look at the problems facing the individual countries that are supplying Japan. Chapter seven reviews the recent long-range forecasts produced by the Japanese government. Chapter eight then presents the author's econometric model of the forestry sector in Japan and presents long-run forecasts. Finally, Chapter nine attempts to draw the threads of the argument together and presents some conclusions.

Chapter one:

An Outline of Japan's Forest Sector

Introduction

The annual consumption of timber in Japan is about 0.9 cubic metres roundwood equivalent (RWE) per capita. This is not an outstanding figure. It is less than the figures for North America and Scandinavia and it is not much different from the figure for Britain. However, given Japan's population, this represents a high level of consumption. Like Britain, the bulk of this is met by imports with only one-fifth of Japan's demand for timber products being domestically supplied. The result is that Japan is one of the world's major importers of timber and timber products. FAO figures show that the value of Japanese imports constitute about 13% of total world trade in forest products, making Japan second only to the USA. If the trade between the USA and Canada is excluded, it easily becomes the world's largest importer of forest products. In certain markets Japan is even more significant than the overall figure of 13% suggests. Japan is the largest importer of industrial roundwood and wood chips. The well-known Japanese preference to import raw material rather than finished products is reflected in the fact that it imports a relatively smaller proportion of all of the sawnwood and wood-based panels traded internationally, although it accounts for a little under a quarter of all of the world's plywood imports.

However, somewhat paradoxically, Japan has abundant timber resources, two-thirds of the country being forested. Not only that, but the forest plantations of Japan, which were largely established in the years immediately after the Second World War, are now reaching maturity. Such is the scale of this maturing timber stock that its felling would have a major impact on world timber markets. To give an idea of the magnitude of this potential supply consider Japan's man-made forests which are almost entirely coniferous. The area of these forests aged between 46 and 55 years is set to increase about sevenfold over the next 20 years. The economic maturity for the species concerned is usually quoted as being about 50 years. If only half of this maturing timber stock were felled the supply would be sufficient to satisfy Japan's present domestic demand and then the entire coniferous log imports of the rest of the world, converting Japan from the world's largest importer of coniferous timber to its major exporter. However, the output from Japan's forests has long been in decline. Price competition from imported timber and rising costs at home have depressed Japan's forest sector making such optimistic projections of output untenable.

Two prognoses for forestry in Japan are possible. In one, domestic supply continues to contract, Japan's forests remain unfelled and Japan's demand for imports continues. In the other, some of the maturing timber is felled, contributing to domestic supply and reducing Japan's appetite for timber imports. This book will examine the issues which surround these contradictory prognoses and offers forecasts of future supply and demand patterns. To begin with, however, this chapter looks at the background to the forest sector and sets the scene for the discussion which follows.

Geographical Background

Japan consists of an archipelago which stretches about 3000 kilometres, from latitude 45° north almost to the Tropic of Cancer at its southernmost islands. Most of its area is taken up by the four main islands of Hokkaido, Honshu, Shikoku and Kyushu, the latter being the most southerly of the four, almost reaching a latitude of 30°.

The total land area of Japan is around 37.77 million hectares, making it about half as big again as the United Kingdom and somewhat smaller than California. About 13% of this area is used for agriculture and around 20% is classified as urban and other land. This leaves a forest area of 25.25 million hectares, of which 23.88 million are classified as 'closed forest'. The overall forest area gives Japan a forest cover of 67%. This is a very high figure for an industrialized country, all the more so when the population density of around 0.3 hectares a person is considered. Only Sweden and Finland can match the forest cover of Japan and they are considerably less densely populated. The high population figure means that the per capita forest cover is only 0.2 hectares, a lot lower than the world average and lower than the figures for many European countries and North America.

It would not be too much of an exaggeration to say that the topography of Japan is characterized by coastal plains which give way to the mountains of the interior with very little rolling countryside. The only exception to this might be Hokkaido which does have areas of rolling landscapes. This impression is backed up by figures which show that only 30% of Japan has a slope of less than 8° (Izumi, 1988). Most of Japan's forests are to be found on the 70% of the land area with slopes greater than 8°. Indeed, for many Japanese the word mountain (*yama*) and forest are virtually interchangeable, a confusing usage in the one or two flat areas of Japan. Here farmers will happily refer to their *yama*, meaning their forest, when surrounded by flat land.

The average rainfall for Japan is 1700 mm with a range from about 1000 mm to 4000 mm. The mountain ranges of the interior of Honshu divide it into two halves and this influences the distribution of precipitation. On the Japan Sea coast precipitation of around 2000 to 3000 mm is typical while on the Pacific coast the figure is typically around 1500 mm. On Kyushu, however, the pattern is reversed and the Pacific coasts of Shikoku and Kyushu experience the greater amount of precipitation. There is a wide range of average annual temperatures, from only about 6°C in Hokkaido to around 16°C in Kyushu and even higher in Okinawa. The low figures experienced in Hokkaido, and to a lesser extent in other areas, mean that a great deal of the precipitation falls as snow. The average minimum temperatures for Hokkaido and the northern prefectures in Honshu fail to pass freezing point. As well as this wide range in the overall annual average temperature, the average monthly temperature in a given location shows wide variation. In Tokyo the average temperature ranges from a low of 4.7°C to a high of 26.7°C. This range of temperature is, from a personal point of view, exacerbated by the humidity figures with summers tending to be uncomfortably humid and winters a little too dry for comfort. These factors are important, not only for tree growth, but also for the traditional housing design in Japan which, more than most factors, influences the nature of timber demand.

Japan lies at the northern limits of the monsoon belt and, with the exception of Hokkaido, experiences a rainy season in early summer. There is also a 'typhoon season' in September, but most of the country is too far north to suffer the worst effects. Even so, given the steepness of the terrain, the rivers in Japan can change quite rapidly in periods of prolonged rainfall. Loss of life and destruction of housing in flooding and land slips are not uncommon in the 'typhoon season'.

Japanese Forest Types and Classifications

Here three ways of classifying Japan's forests will be looked at. First, the natural forest types of Japan will be briefly examined. The classification by ownership and means of regeneration will then be discussed. These last two means of describing Japan's forests are of more importance from the point of view of timber output so the discussion of them will be continued in Chapter three.

Natural forest types

The mountainous nature of Japan and the fact that, being on the western Pacific coast, it does not experience the moderating maritime influences enjoyed by Europe, mean that the range of natural forest types is large. Here four main forest types will be considered and their distribution is shown in Fig. 1.1.

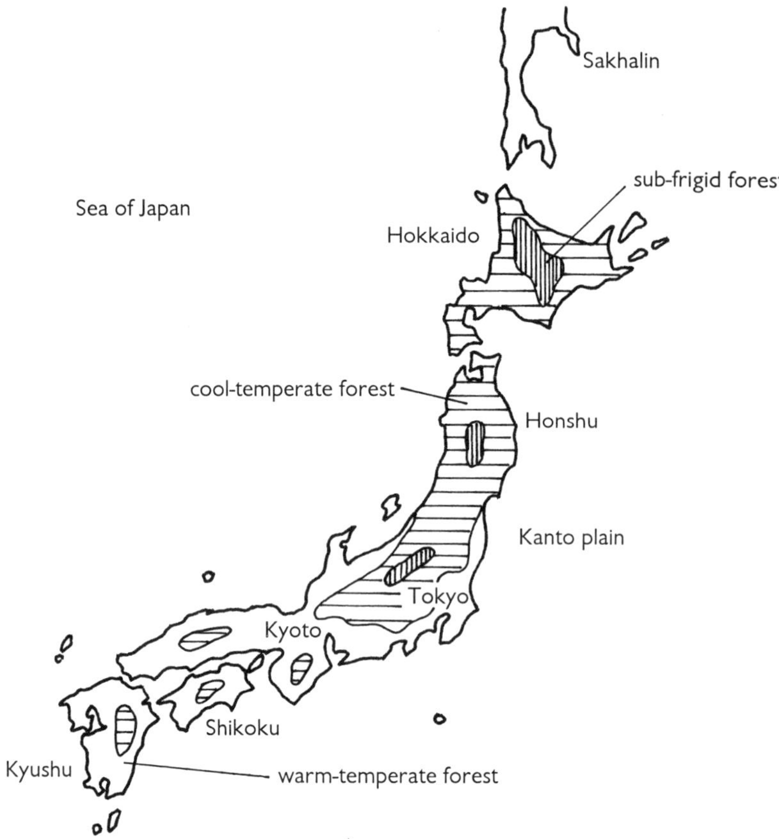

Fig. 1.1. The main Japanese forest types

In the south of Japan, in the Ryukyu islands of which Okinawa is the largest and where the average temperature is about 20°C, subtropical forests can be found. The area of these forests is small and they are not of any great economic importance. However, some areas of mangrove forests still exist and the timber from the forests is used in local markets. The warm-temperate forest zone is characterized by evergreen broad-leaved forests. Such forests occur extensively in Kyushu, Shikoku and the southern and coastal areas of Honshu. Their main constituents are oaks and *Castanopsis* spp. There has been significant planting in this zone and so a number of man-made coniferous forests can also be found.

Stretching northward from the central part of Honshu to the southern parts of Hokkaido is the cool-temperate forest zone. Here average temperatures range from 6°C to 13°C. The main naturally occurring species are beeches, oaks and maples. Sometimes the important coniferous species dominate instead of the broad-leaved trees. However, bamboo is often found on the forest floor in this zone and this causes problems in regeneration.

Finally, in the sub-alpine zones of Honshu and Shikoku and in the mountainous areas of Hokkaido where average annual temperatures are below 6°C, is a sub-frigid forest zone comprising coniferous species such as spruces and firs. Again, bamboo can cause problems.

Ownership of forests

The Japanese divide their forests into national (*kokuyuurin*) or private and public (*minyuurin*). A little under one-third of Japan's forests is nationally owned, the rest being in 'private or public' hands, which includes local communities and prefectures. The figures for 1995 are given in Table 1.1.

Table 1.1. The ownership of Japan's forests

	Area		Growing stock	
	(000 ha)	%	(000 m^3)	%
National (*kokuyuurin*)	7,844	31.2	912,070	26.2
Kokuyuurin	7,647	30.4	892,460	25.6
Other	197	0.8	19,610	0.6
Private & public (*minyuurin*)	17,302	68.8	2,571,160	73.8
Shiyuurin	14,572	57.9	2,212,100	63.5
Shichousonzaisanku	1,534	6.1	210,160	6.0
Todoufuken	1,196	4.8	148,900	4.3
Total	25,146		3,483,230	

source: Forestry White Paper (1997)

The figures in Table 1.1 contain some areas of empty land. In total this is 1214 thousand hectares divided roughly equally between national and private forests. They also include some bamboo stands, these being 152 thousand hectares in all and entirely in private hands. The growing stock of private forests averages 149 cubic metres per hectare whilst for national forests the figure is only 116.

National forests are further subdivided into two groups. The largest group, also rather confusingly referred to as *kokuyuurin*, is under the control of the Forestry Agency which is part of the Ministry of Agriculture, Forestry and Fisheries (MAFF). The other grouping is forests that are owned and managed by other organs of central government. At the time of writing, there were discussions under way about government reorganization in Japan and the idea was being aired of splitting the Forest Agency and placing part of it into a newly reorganized Environment Ministry.

A large part of the national forests is located in Hokkaido, where slightly over 3 million hectares are found. Looked at another way, about 56% of Hokkaido's forests are nationally owned, although a fairly large proportion of this is not closed forest. Given the comparatively recent development of Hokkaido, the pattern of private ownership seen in the rest of Japan was slower to develop and so the predominance of national forests is not surprising. The proportion of national forests tends to diminish towards the south. In northern Honshu the figure is a little under one-half whilst on southern Honshu the proportion drops to less than one-tenth. In Kyushu the figure is generally lower than the overall average. In isolated prefectures large areas of national forest can be found, especially in national parks. For example, in the mountainous areas of Nagano in central Honshu the proportion of national forest is over one-quarter.

By far the largest part of private and public forests are privately owned (*shiyuurin*), mainly by individuals but also by companies, temples and other private organizations. The other categories of private and public forests are either in the hands of prefectures (*todoufuken*) or are owned by cities, towns and villages (*shichousonzaisanku*).

The largest individual category within the *shiyuurin* group consisted, at the last Forestry Census in 1990, of over 2.5 million 'forest families' or *rinka*. Their average holding was only 2.7 hectares and nearly 1.5 million held less than a single hectare of forest. The next most important category, in terms of area of forest, was companies. These held just over 1.5 million hectares of forest with an average holding of around 35 hectares. Some of these forests, especially those in Hokkaido where such company forests are common, are used explicitly for forestry purposes, being owned by pulp and paper companies or sawmilling concerns. However, in the recent property boom, a number of such forests were purchased in the hope of converting them to other uses.

Natural and man-made forests

The Japanese also divide their forests into man-made (*jinkourin*) and natural (*tennenrin*), the distinction being the method of regeneration. Man-made forests are those reproduced by some form of planting whilst natural forests have been regenerated using natural seeding. Of the total forest area about 41% is man-made.

Table 1.2 gives the breakdown of man-made and natural forests in terms of ownership. The figures show that national forests are predominantly natural whilst the split in private forests is far more even. The growing stock of the man-made forests exhibits a wide divergence between the private and national categories, with the former showing a figure of 201 cubic metres per hectare whilst national forests manage only 119. In the natural forests the ranking is reversed with the national forests showing the higher stocking of 131 cubic metres per hectare.

Overall, man-made forests contain 54% of the growing stock whilst accounting for only 41% of the total area. More importantly, man-made forests contain almost all of the valuable coniferous forests whilst the natural forests are composed predominantly of broad-leaved trees with a lower value. It is in the cool-temperate forest zone that man-made coniferous forests are most widely spread although as a very rough rule of thumb, the proportion of man-made

Table 1.2. Natural and man-made forests

	Natural forests		Man-made forests	
	Area (000 ha)	Growing stock (000 m^3)	Area (000 ha)	Growing stock (000 m^3)
National forest	4,738	618,710	2,446	292,230
Private forest	8,644	971,310	7,952	1,599,760
Total	13,382	1,590,020	10,398	1,891,990

Unforested and bamboo areas excluded so figures do not equal those in Table 1.1.
source: adapted from Forestry White Paper (1997)

forests increases towards the south. They are of major importance for the supply of timber to the Japanese economy and the post-war years have been characterized by the conversion of natural forests in which broad-leaved species were dominant to man-made coniferous monocultures.

The Major Industrial Species

It might be useful to list the main industrial tree species found in Japan. Their Japanese and botanical names are shown in Table 1.3 with, where appropriate, their English names. The Japanese names will be used in the following discussion as, in most Japanese publications, these rather than the botanical ones are used.

Table 1.3. The main species used for industrial purposes in Japan

Japanese name	English name	Botanical name
Sugi	Japanese cedar	*Cryptomeria japonica*
Hinoki	Japanese cypress	*Chamaecyparis obtusa*
Akamatsu	Red pine	*Pinus densiflora*
Kuromatsu	Black pine	*Pinus thunbergii*
Karamatsu	Japanese larch	*Larix leptolepis*
Ezomatsu		*Picea jezoensis*
Todomatsu		*Abies sachalinensis*
Momi		*Abies firma*
Tsuga		*Tsuga sieboldii*
Nara	Oak	*Quercus* spp.
Buna	Beech	*Fagus crenata*

Most of the species in Table 1.3 are indigenous to Japan. Japanese forestry, unlike British, is not based on exotic species. However, few of the species are dominant in natural forests. Most of the species are to be found naturally in the cool-temperate forest zone, although karamatsu and todomatsu make up the bulk of the sub-frigid forests in Hokkaido where they are found in a natural forest setting rather than plantations. Forests of momi, nara and tsuga are also seen in the warm-temperate forest zone.

As in most countries, there exist both national and local markets for timber. Many species are used in the local markets, for example timber from the subtropical forests are used for local turnery. However, it is the national markets that are of significance here and, as would be expected, it is the coniferous species which are of major importance. Furthermore, the species which form the backbone of these markets are the planted conifers. Species that dominate the natural forests tend to be of lower value.

Figures from the 1997 Forestry White Paper (referring to the year 1996) for the volumes of timber felled in Japan show that just over 80% of domestic timber supply was accounted for by coniferous species. Of this, the lion's share was taken by sugi, which accounted for just over 40% of total domestic log production. Next in importance came hinoki, accounting for 13%. The other coniferous species taken together accounted for 27% of domestic timber output. The hardwood species listed in Table 1.3 only contributed 2% to domestic supply. The remaining 18% or so consisted of hardwood species which the Japanese refer to as 'others'.

It has already been noted that coniferous species are chiefly to be found in man-made forests and so this dominance of the conifers, and especially sugi and hinoki, in the domestic market means that man-made forests are of great importance. The bias towards planting coniferous trees is seen in the species composition of man-made forests which could not be more distinct from that of natural ones. The figures from the latest 10-yearly Japanese forestry census of 1990, are given in Table 1.4.

Characteristics of the main species

A brief description of the main species might be useful as it relates to their end-uses and market demand. Sugi is a rapidly growing conifer which has been extensively planted in Japan. It is a shade-intolerant species and so a great deal of

Table 1.4. The species composition of Japan's forests

	Natural forests		Man-made forests	
Species	Area (000 ha)	%	Area (000 ha)	%
Sugi	0	0	4,508	44.1
Hinoki	0	0	2,396	23.5
Akamatsu & kuromatsu	1,090	8.1	1,024	10.0
Karamatsu			1,077	10.5
Other conifers	1,263	9.4	1,008	9.9
Buna & nara	1,072	8.0	61	0.6
Other hardwoods	9,954	74.4	140	1.4
Total	13,379		10,214	

Figures refer to 1990 and so do not quite equal those in Table 1.1.
source: Akaha (1992)

tending is optimum, especially thinning and early weeding. Its timber is very smooth when sawn and the grain is straight. It is a widely distributed species, occurring mainly in the area from north-west Honshu to south-west Kyushu. Okinawa is the only area of Japan in which it is not found. Although sometimes found naturally in the cool-temperate forests, sugi is found almost entirely in man-made forests as is shown in Table 1.4. The most common approach to its cultivation is to establish plantations with densities of 3000 to 4000 stems per hectare. Seedlings or rooted cutting are used. Three or four thinnings take the stem density to 600 to 800 by the usual rotation age of 40 to 60 years.

Hinoki is again a widely-planted species, mostly on the middle and tops of slopes. It is a more shade-tolerant tree than sugi when young but, as it grows, thinning becomes necessary to maintain optimum growth. It is somewhat slower growing than sugi. It has a very fine texture and lustre. Its white colour is highly valued and, in the past, the use of the species was restricted in many areas of the country, being reserved for the houses of the nobility and temples. It can also be a very durable timber. *Houryuu-ji*, a temple in Nara prefecture, is reputed to be the oldest wooden building in the world, being built about 1300 years ago. It is constructed largely of hinoki.

Hinoki is found in man-made forests. The only places in Japan where it does not occur are Hokkaido, the northern prefectures of Aomori and Akita and on the Ryukyu islands. As a general rule, it is planted 'further up' the mountainside than sugi as it prefers drier soil conditions. Indeed, if the site conditions are too favourable hinoki tends to suffer from rot and produces larger than usual quantities of resin. This is a major problem as the price premium that hinoki commands is dependent upon its visual characteristics and any visible resin drastically reduces its value. Initial planting densities are usually somewhat lower than those for sugi and frequent thinning and pruning are required if good quality saw timber is to be produced.

Although it is hard to generalize about forest growth in man-made sugi and hinoki plantations, on average sites in central Japan sugi achieves growth rates which give maximum mean annual increments of around 12 cubic metres per hectare a year. The growth rates of hinoki are slightly lower than this. Both sugi and hinoki are somewhat faster growing than similar trees in the British context, reaching their points of maximum mean annual increment at ages between 35 and 50 years. Again, as a broad generalization, the volume production of well-managed stands of sugi is somewhat higher than stands of Douglas fir or Western hemlock on the American West Coast until the age of around 50 when the growth rate of sugi slows and the firs and hemlocks overtake the Japanese species (Kato, 1981).

The other coniferous species are usually deemed to be of a lower quality than sugi and hinoki. Todomatsu is exclusively found in Hokkaido and is fast growing. However, there are problems with windblow. It is a weak timber and not really suitable for traditional construction purposes where strength is needed.

Karamatsu has good frost resistance and its initial growth in Japanese conditions is rapid and so it was expected to be the favoured species in areas too severe for sugi or hinoki. It grows well in the area from the warm-temperate zone to the sub-frigid forest zone in Honshu and it is mainly found in the upland areas of Honshu and in Hokkaido. However, in pure plantations there has been some pest damage and also problems with the timber quality. The colouring shows a distinct boundary between the white heartwood and brown heartwood and the wood is very resinous.

Both akamatsu and kuromatsu are used for coastal protection forests. However, there have been problems with pinewood nematodes carried on pine sawyer beetles and major aerial spraying has been undertaken to reduce the damage. Ezomatsu is found only in Hokkaido. It is slow growing and the timber from the old-growth natural forests is valued for high quality items, one of its special uses being for piano sounding boards. The lower quality timber goes for pulp.

The end-uses of Japanese timber

The brief comments about the qualities of the species used in Japanese forestry are borne out by their end-uses. The Japanese do not publish very detailed end-use figures but the data for 1996 are shown in Table 1.5. The figures for 'chips' should be interpreted as meaning chips for future pulping.

Table 1.5. The end-uses of Japanese timber in 1996 (all figures are percentages)

	Sawnwood	Plywood	Pulp	Chips	Other uses
Sugi	96.3	*	*	2.5	1.2
Hinoki	97.0			1.5	1.5
Akamatsu & kuromatsu	54.0		19.3	22.9	3.8
Karamatsu	72.7	2.6		21.1	3.6
Ezomatsu & todomatsu	61.4	*	34.0	3.4	1.1
Momi & tsuga	83.3			12.7	4.0
Nara & buna	57.3	12.2		30.3	*
Other hardwoods	13.0	3.0	1.3	80.0	2.6

* Small level of usage recorded but less than one-tenth of 1%.
source: adapted from *Mokuzai jyukyuu houkokusho* (1997)

The table shows that almost all of the sugi harvested is used in the sawmilling industry. Although not shown in the table, most ends up in housing construction and furniture. Housing in Japan uses a great amount of timber. The basic items in traditional architecture are pillars and beams, the former being either 10.5 cm or 12 cm squares and local sawmills tend to specialize in their production. The nature of the housing in Japan is that these pillars are often visible. Even in modern concrete apartment blocks Japanese style rooms with visible timber features are usual. The visual qualities of the species used in housing are, therefore, very important with clear, smooth, white timber the most desirable. Hinoki is also almost entirely used for sawmilling. Its history as a major species in temple and palace construction and its appearance enables hinoki to command a premium price compared to sugi, selling at about double the price. It is used mainly in housing construction, furniture and small fixtures, these always being of a higher quality. Indeed, there is something of a feeling that hinoki is a species which is associated with wealth and that its inappropriate use might be looked down upon as somewhat *nouveau riche*.

This pattern of usage with sugi and hinoki being used almost exclusively for sawn timber is not surprising when one considers the relative prices in the domestic market for saw logs and the price of pulpwood or chips. Typically, the price of logs going as pulpwood has been about half to a third that of sugi logs for sawmilling and this, in its turn, has been about half the price of hinoki logs. As these are the favoured species in building, there is an incentive to manage them in a way so as to produce the more valuable building timber. However, the high figures for the proportion of sugi and hinoki used for sawnwood are also indicative of the fact that little of the waste produced is used in, for example, chip production. The usual practice is to cut the squares for use as pillars in housing, cut other dimensions for housing use or other fixtures and then discard or burn the waste.

The timber of the other coniferous species tends to be of a coarser quality and so goes for lower quality construction purposes and pulping. The pine and spruce timber which is used for sawnwood typically produces a product which has a slightly lower value than that of sugi. For example, 10.5 cm square pillars of ezomatsu typically sell for a price about one-tenth to one-fifth lower than the equivalent grade of sugi. Todomatsu is used for packaging and is one of the best timbers for pulping.

The combined akamatsu and kuromatsu figures in Table 1.5 hide the bias for akamatsu to be used in buildings. About four-fifths of the akamatsu sawn timber is used in building or civil engineering uses whilst the rest is used in packaging. The timber going for house construction does not usually compete with sugi and hinoki in the market for decorative elements in construction although it is used in some areas, being popular in the Kanto region and Chuugoku. Its high shear strength makes it useful for items such as beams. The kuromatsu sawnwood is mostly used in packaging.

The nature of karamatsu means that it is less popular in housing construction although its elastic qualities make it useful in various areas such as vehicle construction. It is durable and strong so it is valuable for the parts of housing that are not visible, its visual characteristics and rough texture mean that it is not popular for the visible elements of Japanese housing. A large proportion of the sawnwood goes for packaging and for transport in pallet construction (Ishii and Arai, 1986).

The natural broad-leaved forests in Hokkaido, where growth is slow and so timber quality high, have high value and are used in housing for such items as door frames and windows.

The use of hardwood chips, almost entirely used in the production of paper, is of significance as, in the post-war history of Japanese forestry, the broad-leaved forests have been cleared to make way for the more valuable coniferous species. This trend, which gained momentum after the collapse of the market for charcoal and wood for fuel, is one of the reasons that almost all of the man-made forests in Japan are coniferous in nature. Also, this movement to coniferous species was seen by the Japanese at the time to be definitely desirable. The establishment of coniferous plantations at the expense of broad-leaved forests is referred to in the statistics as *kakudai zourin*, which roughly translated means afforestation to expand (rather than convert) the area of forest.

Overall, Table 1.5 shows that Japanese forestry is mainly oriented to the production of sawnwood, at least as far as the more valuable coniferous timber is concerned. The overall consumption of timber, in volume measure, shows that 72% of the domestic supply of timber was used in the sawmilling industry and, for the coniferous species, this figure was 85%. Over recent years, however, the use of conifers in plywood production has started to increase, in part as a result of the increasing expense of tropical timber. The present figure of around 0.25% of coniferous roundwood entering the plywood industry represents a 'major' increase over the previous figures.

The pattern of species and ownership distribution and end-uses discussed above is a snapshot of the present situation. As has been alluded to, it is in large part a result of the history of Japanese forestry and so it might be useful to look briefly at the main features of the development of Japanese forestry.

A Brief History of Japanese Forestry

The early years

The earliest recorded boom in timber demand, resulting from the need for construction, occurred in the period AD 500 to 900 during the Yamato court. This dynasty succeeded in unifying large areas of the country and started a programme of temple and palace building, establishing many of the temples in the Kyoto and Nara areas, including *Houryuu-ji* mentioned earlier. Unlike the western experience, the growing wealth in Japan was not reflected in a move from timber housing to more durable materials such as stone. As a result, during this period a shortage of large-sized timber in the Kyoto and Nara areas was reported. The response was a felling ban in some forests and the expansion of logging into other areas.

After this time the country suffered from a prolonged period of civil wars and it was not until 1585 that Japan was reunited under what was to become the Tokugawa Shogunate. One of the notable features of this era, which made Japan distinct from most other countries, was the policy of *sakoku*. This was a policy followed by the Shogunate which more-or-less closed the country to international trade. It was established, partly at least, in an attempt to exclude Christianity from the country. However, an indirect effect was that for the years before the middle of the nineteenth century, when the great cities of Japan were being established and maintained, the demand for timber for construction had to be satisfied by domestic production.

During the Tokugawa Shogunate the urbanization of Japan proceeded apace and the population reached 30 million by the beginning of the eighteenth century. The demand for timber which this created was very large. Also, the production of rice to feed the urban population meant that a large area of forest was needed to provide green manure for fertilizer use. The relatively low level of meat consumption meant that animal manure was not available in the quantities which were produced in other countries. Most of the supply of green manure originated in the communal forests which were managed under a system of mutual agreement by community members (Funakoshi, 1981). This need for green manure and the urban population's demand for charcoal meant that around 12 million hectares of accessible forest (about 32% of the total land area) were needed for these non-timber products (Izumi, 1988).

The earliest demands for timber under the policy of *sakoku* were satisfied by what has been called a command system, with the rulers of the early Tokugawa Shogunate requisitioning timber for their large building projects. However, gradually an entrepreneurial system evolved with merchants, brokers and the beginnings of the present complicated internal distribution system emerging (Totman, 1987). By the end of the seventeenth century Osaka, and then Tokyo, and emerged as nation-wide timber markets. This period also represented the beginning of private forestry, with forests being regenerated by land owners with a view to future profits. The south-west of the country was the site of most of this early forestry and it is in these areas that good quality timber can still be found today under the traditional management systems first developed in this period.

In the early days, as in most countries, the Shogunate and feudal clans cut the natural stands with little regard for explicit regeneration activities although the nature of Japan's topography meant that such regeneration was soon needed. But by the end of the period replanting with sugi and hinoki was being undertaken by ordinary landowners in the areas with good links to the big cities. It was at this time that large areas of man-made forests were produced, especially around Tokyo, Hamamatsu, Kiso, Nara and Takamatsu. In those days the economics were such that forestry could compete with agriculture and much of the planting was done on land which could, otherwise, have supported agricultural crops. This is in contrast with recent planting which has been mostly on the steeper slopes.

As local timber shortages emerged the Shogunate turned its attention to forest preservation rather than exploitation and went as far as to ban the felling of certain species (Funakoshi, 1988). This policy was used not only for forest preservation but also to earn revenue for the local governments. For example, in the Kiso area in central Japan the policy was known as 'five trees' or *goboku* because five species were reserved for use by the local government. These species were the five most desirable ones, including hinoki and sugi, and they were only to be used for construction of the residences of the local nobility and for temples. Such a policy, with varying species, was instituted around the country and the effects of it are still felt in the demand for timber today, especially in the price of hinoki relative to other species.

The policy of *sakoku* lasted for about 200 years but began to weaken in the nineteenth century and was finally overturned with the fall of the Tokugawa Shogunate in 1868 and the so-called Meiji Restoration.

The years after the Meiji Restoration

The absence of a large navy and mercantilist stage in the development of Japan, meant that there was no large demand for ship timber which guided a lot of the forest policy in the west. However, the rapid development of a railway system in the Meiji era (1868-1911) meant that additional demands on forests other than housing and green manure rapidly developed. Also, the associated reduction in transport costs meant that the extensive margin of forestry could expand.

As well as the development of a railway system, the pulp and paper industry was rapidly developed as one of the means of industrializing the country. As a result of these two developments, timber shortages in some areas began to develop. Towards the end of the Meiji era the forest industries moved into the northern island of Hokkaido and the new colonies of the Sakhalin islands. This expansion had two aims, one was to reduce the timber shortages referred to and the other, somewhat paradoxically maybe, was to establish an export-based forest industry (Narita, 1997). It was in the middle of the Meiji era that the first real planting boom of modern times occurred and this was associated with the development of large-scale ownership. This was encouraged by the breaking up of communal forest and the establishment of titles to such forests as they became either nationally or privately owned.

The Taisho era (1912-1926) witnessed the beginnings of the dependence of Japan on imported timber. With the rapidly growing urban population, demand for timber grew as housing was still, for the most part, wooden. Imports were, as today, mainly from North America, Russia and tropical timber from Southeast Asia. Of course, with Japan's entry into the Second World War, these imports were curtailed and the demand for timber had to be met from domestic forests.

This pattern of demand and the corresponding import levels are shown in Fig. 1.2. The diagram shows that, despite the imports, domestic supply predominated in the years between the First and Second World Wars. Domestic supply came mainly from private forests, with their output rising to over 21 million cubic metres by 1939. Output from national forests also increased but not by the same amount, rising from 2.5 million to 5.5 million cubic metres over the period shown in the graph.

The area of reforestation undertaken by the private sector did not show the

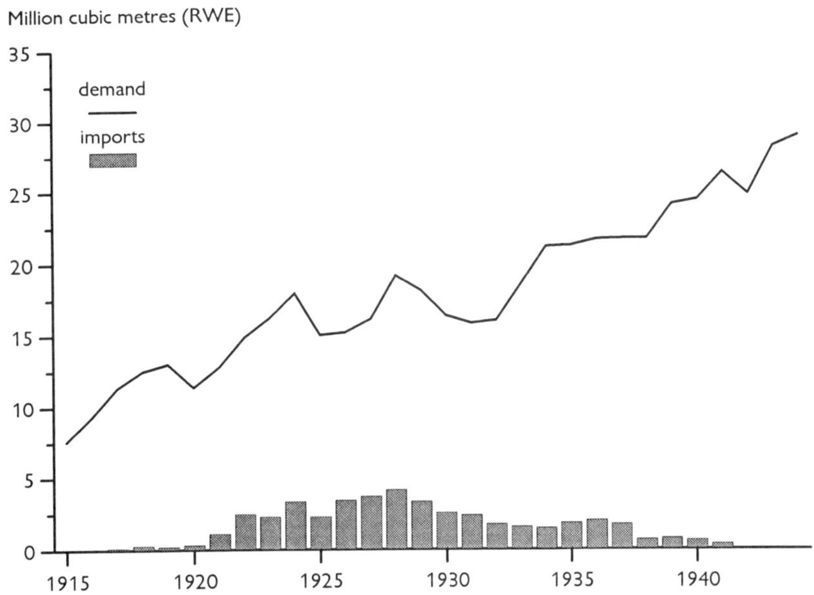

Fig. 1.2. Demand for timber and imports between the wars

same increase as the output figures did. The level of reforestation was only 84.3 thousand hectares in 1940, just over the level registered in 1915, although there had been a slump in the intervening years. National forests, however, showed a pattern of decreasing planting with the area planted falling from 35 thousand hectares in 1915 to only 9 thousand in 1940. This pattern of increased felling and declining planting characterized the later years of the Second World War and set the scene for post-war forest policy.

The years after the Second World War

After the war there was a large number of people returning to Japan, some as a result of the demobilization of the armed forces and some from the former colonies of Manchuria. As Japan was, in terms of population distribution at least, largely an agricultural country the vast majority of these people returned to rural areas and it was in this sector that large-scale employment had, in the short term at least, to be generated. The concomitant fear of rural unrest, and the idea that such unrest had previously contributed to Japanese aggression, meant that a land reform package was developed by the occupation forces. According to this package all land owned by absentee landlords could be compulsorily purchased. The maximum area which resident landlords or owner-occupiers were allowed to hold was limited to around 12 hectares in Hokkaido and to only a quarter of that in other areas. Any areas over this were purchased and then resold, on favourable terms, to existing tenants (Beasley, 1982).

Although these laws were not applied to forest land, the effect was to create a large number of very small-scale farmer/foresters. This generated a pattern of land holdings which was, in the absence of active markets in forest land, to mitigate against the ability of the Japanese forest sector to compete with the more industrial style of forest plantations in other countries which were to come to dominate the international market.

As a result of the war, with its large demand for timber and effects on the availability of rural labour, the area of felled forest which had not been replanted reached 2 million hectares, 1.6 million of which were in private hands (Sugahara and Ouchi, 1988). The topography and climate of Japan meant that this situation, if left unremedied, would have had serious environmental consequences. Under the terms of the First Basic Forest Law, passed in 1951, prefectural governments were made responsible for overseeing the planning of private forestry, although the individual owner was still responsible for actual operations. The Law contained regulations which forbade the cutting of timber below the recognized felling age without specific permission. This felling age was defined as the age of maximum mean annual increment and was usually around 35 to 45 years (Handa, 1988). There were no limits placed on cutting timber over the rotation age but, after the war, the remaining accessible forest in this category was so small that no real increase in supply resulted from this concession.

Also, the bombing raids on the mainland had caused a high level of destruction, especially given the wooden nature of most of Japanese housing. At the same time, the urban population needed charcoal for its heating. So, simultaneously, there was a high demand for forest products and a low supply potential.

Not surprisingly, the demand for firewood fell rapidly in the reconstruction period as heating needs switched to more modern sources of energy, but at the time it was satisfied by the felling of broad-leaved natural forests. The building timber requirements were largely met by coniferous timber output and, in response to the apparently favourable market conditions that this created, the felling of timber for firewood was usually associated with the conversion of forests from natural broad-leaved ones to man-made coniferous ones.

During the 1960s Japan's economy grew rapidly and the demand for timber grew along with it. Part of this was the increase in the demand for construction timber but also the increase in the demand for paper products and panels meant that the overall demand level increased with the growth of Gross Domestic Product (GDP). Demand for saw logs and lumber increased by around 5% a year, pulp demand by around 8% and panel products, almost entirely plywood, by 15%. In 1973 the first 'oil shock' was felt by Japan and after that the economy entered what the Japanese like to call a period of 'stable growth'. The effect on timber demand, however, was somewhat more profound with demand dropping and then hardly growing at all in the subsequent years.

The effect of the oil shock was different for different product categories. The worst affected was the demand for saw logs and lumber, which did not grow at all in the ensuing years. On the other hand, the demand for pulp and panel products grew somewhat, although not at the rate which they had shown before the oil shock. A final period of rapid growth was observed in the late 1980s, during the so-called 'bubble economy', with growth in all categories but mostly in the pulpwood and panel products areas.

The pattern of demand just discussed is shown in Fig. 1.3 and it is a pattern that is reproduced in many sectors of the Japanese economy, especially in housing which is important for the rest of the discussion. A rapid expansion is followed by a break in the growth and then by either slow growth or stability.

The thinner line in Fig. 1.3 shows the self-sufficiency ratio defined by dividing the domestic supply of timber by the overall level of demand. This is a much-quoted statistic in Japanese forestry, the so-called *jikyuuritsu*. It varies for the different elements of timber demand. Notably, the demand of pulpwood and chips, and veneer logs for plywood manufacture is much more dependent on imported supply whilst the demand for saw logs is far more likely to be met by domestic output. However, the picture presented in the diagram is one of a steady decline in overall self-sufficiency from its immediate post-war level of nearly 100%.

After the war the Japanese government exerted strong control over foreign exchange. Hard currency was not available to finance the import of timber, the money being used to purchase imports of what were deemed to be more

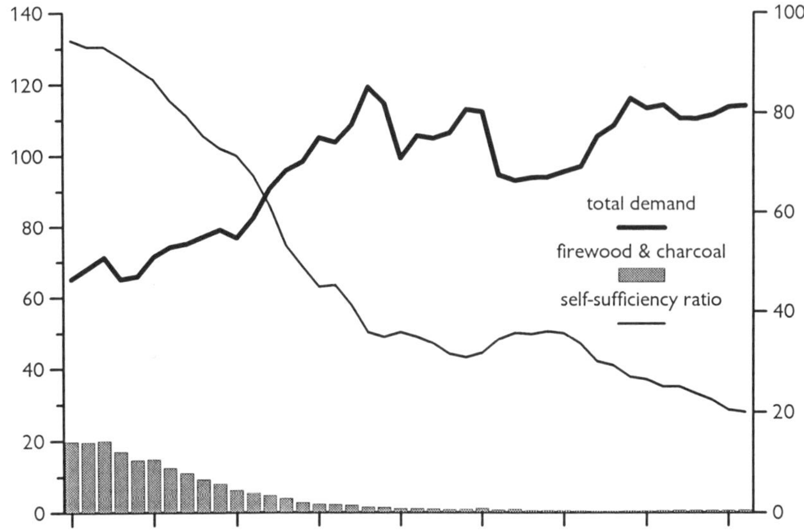

Fig. 1.3. The post-war demand for timber

important commodities. Also, the infrastructure was not really in place to handle large-scale timber imports. There was not only an absence of portside processing plants but also a shortage of tonnage actually to transport timber. As a result, these years were characterized by a high level of demand being satisfied almost entirely by domestic supply. The result of this self-imposed import 'ban' was an increase in the real price of coniferous timber. Figure 1.4 shows the movements of the price of sugi logs and the wholesale price index (WPI) over the post-war period. The series have been rebased to 100 at various stages to make the graph easier to interpret.

During the period until 1960 the real price of timber, especially coniferous timber, rose rapidly. The increase in the price of conifers was one of the factors encouraging the conversion of broad-leaved natural forests to coniferous ones. However, the increasing timber price was causing problems and so in 1961 the so-called 'Measures to Stabilize Timber Supplies' were introduced. The foreign exchange restrictions which were applied to timber imports were abolished in 1959 and this freed up money for the import of timber (Handa, 1988). The 'Measures' built on this and encouraged the development of an infrastructure designed to handle and process imported logs and, to a lesser extent, sawn products.

The result of all this was that the import of coniferous timber in direct competition with domestically produced timber began in earnest and the self-sufficiency ratio began to fall. In the year that saw the introduction of the

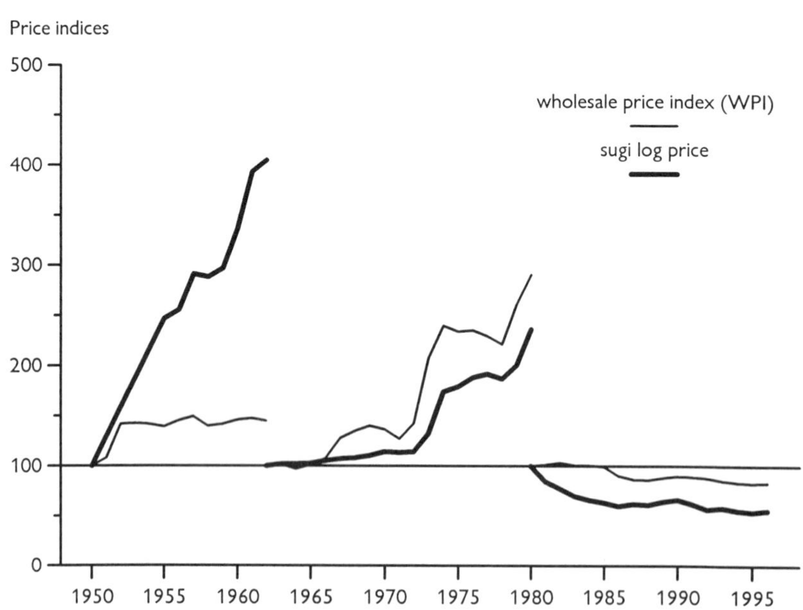

Fig. 1.4. Price movements in Japanese forestry

'Measures', the imports of timber from North America quadrupled and those from the former USSR increased by nearly 50%. The overall effect on the self-sufficiency ratio was somewhat dampened by the fact that Japan was already importing about 90% of the timber it needed for plywood production. But in the saw logs and sawnwood area the self-sufficiency ratio stood at 89.5% in 1961 after having fallen from 99.4% in 1955. However, 10 years after the opening up of the timber market in Japan the self-sufficiency ratio for saw logs and sawnwood stood at only a little over 44%.

The measures were, to a certain extent, opposed by the forest owners but to no avail. The reduction in the price of important building materials was seen as being of more importance than maintaining rural employment, a balance of power between the Ministries of Agriculture and Construction which seems to persist to this day.

The effect on the price of timber was the required one. In the period from 1961 to 1968 the pace of increase of the real price of timber slowed and after that, until the 1980s, it more-or-less followed the WPI. In the more recent past, the high value of the yen and the increase in the penetration of foreign suppliers into the Japanese market has depressed the timber price so that, in real terms, it is not keeping pace with the WPI. The nominal log price is now at almost half the peak levels recorded in 1980.

Of course, this has caused problems for Japanese forests for, although the price of timber was not growing, wage rates were. For private forestry this created problems as, being an adjunct of the small-scale agricultural sector, forests were often felled and planted and the timber transported by family members during the rural 'off-season'. However, the growth of the economy meant that other, better-paid jobs were becoming available and so the underemployment which had characterized the rural economy in the early post-war years disappeared. Figures from the Forestry Census show that in 1960 only 8.1% of forest owners holding in excess of one hectare claimed to obtain most of their income from outside agriculture and forestry. By the latest census in 1990, however, this figure had risen to 53.3% (Akaha, 1992). Not only have opportunities for part-time and full-time work become available for those who remain in rural areas but the relative desirability of work in the towns and cities has increased dramatically. Many of the children of farmers have, after a college education, moved to the city to work for the large companies, leaving their parents to manage the forests.

The overall effect is that the rural sectors of Japan are characterized by declining and ageing populations with the concomitant gradual closures of services, schools, etc. The result of this is that, not only has the total number employed in forestry decreased from 519,000 in 1960 to 108,000 in 1990 but, more significantly, the age structure has worsened dramatically.

Figure 1.5 shows the proportions of three age groups of workers engaged in forestry (technically those who had, in the 9 months previous to the survey, worked mainly in forestry for a period of at least one week). The rapidly increasing proportion of older workers is immediately apparent. Of course, Japan as a whole is an ageing country, a fact that is troubling policy makers. But, while the proportion of those involved in forestry who are aged over 50 has increased from 22% to 68% in the period covered in the graph, in the working population as a whole the increase has only been from 21% to 31% (Handa, 1990; *Zenkoku shinrin kumiai rengoukai*, 1991).

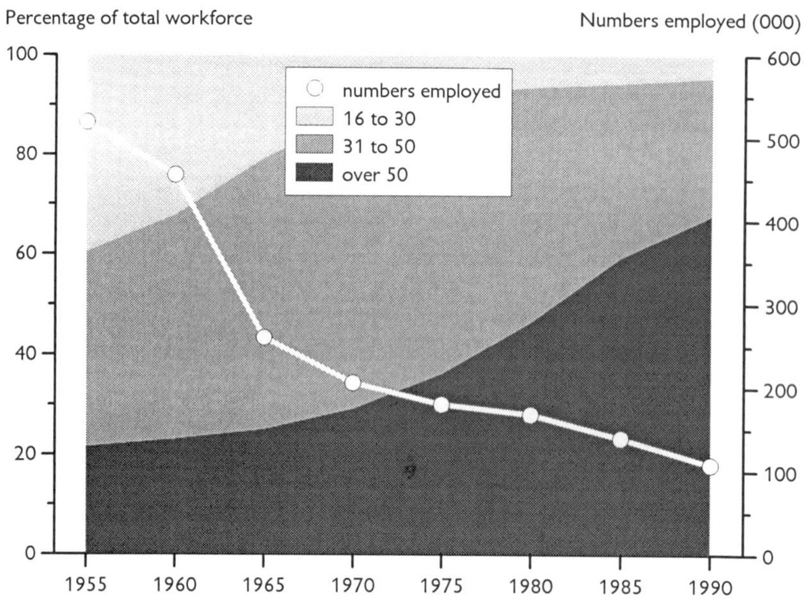

Fig. 1.5. The age structure of forest workers

Similarly, the productivity increases which were being created in other areas of industry were not so easily repeatable in forestry. The terrain problems meant that the high productivity mechanization of logging was not possible and so the profitability of forestry declined at the time that international competition was increasing. Not only that but the floating of the yen in 1973 paved the way to the strengthening of Japan's currency in the late 1970s and 1980s, making the comparative disadvantage of Japan's forest sector even more pronounced.

However, the processing of timber which had been harvested under more favourable circumstances was an area where Japan could invest and, in the post-war years after the 'Measures', the establishment of an import-based timber processing sector proceeded apace. Already there was a plywood and paper industry in Japan but, in sawmilling, the construction of a coastal, import-based sawmilling industry meant that imported logs could be used to produce lumber in direct competition with the output from sawmills using domestic timber. This trend is shown in Fig. 1.6 where the numbers of sawmills using only domestic timber and the number of those exclusively using foreign timber is shown. An additional category of mixed mills also exists but is not shown in the diagram.

Along with the decline in the numbers of the sawmills using domestic timber there was an increase in those specializing in foreign timber although these too have declined in number recently with the direct import of sawnwood. Mills using foreign timber are usually favourably located in coastal areas with good port facilities and nearby consumers. Although the number of mills using domestic timber is greater they are a lot smaller. The annual average timber input of the mills is also shown in the diagram. The mills specializing in foreign timber are not only better located to meet the large demand in the urban centres but also are considerably larger, allowing greater economies in production.

Given the problems just outlined it is no surprise to see that the stumpage price of sugi is falling as a proportion of the log price, indicating that the competition in the final products market and the lack of ability to maintain profitability in the sawmilling sector is forcing down the prices which the processing industry is willing to pay for stumpage. In the 1960s the stumpage price was typically around 65% to 70% of the log price. Now the figure is less than half and the stumpage price stands at only ¥10,810 per cubic metre compared with ¥22,707 in 1980.

Also, the increasing age of those working in forestry shown in Fig. 1.5 is being mirrored in the ages of the small-scale builders and carpenters who produce a lot of the wooden housing in Japan. The absence of skilled craftsmen

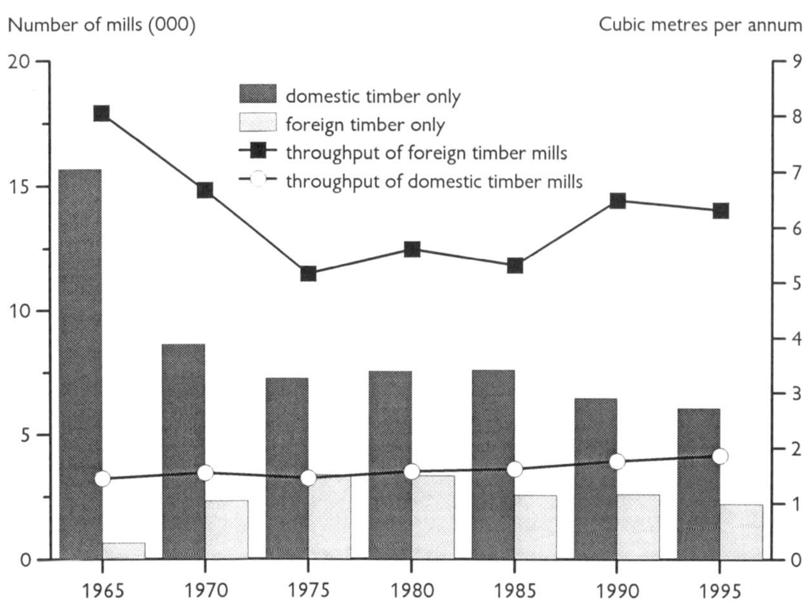

Fig. 1.6. Sawmills in Japan

means that the relative costs of traditional Japanese housing is increasing. At the same time, technical progress in other areas means that substitute raw materials for housing construction have become relatively cheaper compared with wooden materials, and especially the costs of construction and assembly of prefabricated units and large apartment blocks have fallen relatively. This has caused a reduction in the proportion of so-called wooden housing in the Japanese market and the drop in demand for sawn timber.

One of the main areas of demand for the more expensive types of timber from Japan's forests began to decline as the proportion of wooden houses fell. Fashions changed from having traditional rooms with visible, decorative timber which Japan's forests were largely geared to produce, to more western-style rooms where the timber was concealed from view. This opened the way for the use of lower quality timber where price rather than visual considerations were paramount and so increased the comparative advantage of overseas forests.

This brief history of the Japanese forestry sector contains the reasons for the present difficulties in forecasting. Domestic output has been falling for the reasons outlined above. So it is tempting to extrapolate this trend into the future. However, the favourable situation in forestry immediately after the war and the high level of demand led, not only to the felling of domestic forests, but also to planting. As a result a large part of the forests of Japan were planted in the 1950s and 1960s. Given the rotation ages for Japan's forests they would, under normal circumstances, be due for felling in the early years of the next millennium suggesting an increase in domestic supply and reduced import demand. Such is the scale of this volume 'overhang' that a more detailed look at it might be in order.

Man-made and Natural Forests in the Supply of Timber

The striking feature of the figures in Table 1.4, which showed the species composition of Japan's forests, is the fact that the valuable sugi and hinoki timber is almost entirely found in the man-made forests whilst it is not too much of an exaggeration to say that all of the broad-leaved forests are natural. Similarly, as most of the sugi and hinoki timber goes for saw logs it follows that the domestic supply of top quality saw logs is almost entirely satisfied by the output of man-made forests.

Assuming that the output of the various species comes from man-made and natural forests in proportion to the areas in Table 1.4 gives the result that about 83% of the domestic supply of saw logs is supplied by man-made forests. The figure for plywood is 10% and for pulp and chips 18%. These figures are significant because they show that the structure of man-made forests is very important in the saw log market while the structure of natural forests is important for the pulp and paper industry.

Figure 1.7 shows the age class structure of man-made forests in 1990, according to the census of that year (Forestry White Paper, 1992). The age classes used for coniferous species in Japan are 5-year intervals and so age class one represents forests aged 1 to 5 years, age class two forests from 6 to 10 years, and so on. Sugi, hinoki and the other conifers taken separately have age class structures which are similar in form to the overall one for man-made forests shown in Fig. 1.7. Also there is no real difference in the age class structures of private and national forests (Akaha, 1992). The pattern of areas reflects the felling and replanting during and after the Second World War.

To give an idea of the scale of the potential supply that Fig. 1.7 implies consider the following. The area of forest in age classes seven and eight is 2,190,000 hectares. If this were to be felled over the next 10 years, as it progresses into age classes nine and ten (ages 41 to 50), and assuming a final volume of 300 cubic metres a hectare, the average annual timber production would be around 66 million cubic metres. This compares with the present output level of coniferous timber of about 18 million. In the following 10 years the figures become even more remarkable. The 3,360,000 million hectares of forest in age classes five and six would, under the same assumptions, give rise to an annual average output of just over 100 million cubic metres.

Of course, it would be a foolhardy person who forecast such a level of felling. This output would be way beyond the present capacity of the harvesting and processing industries. However, if the weaker assumption is made that felling behaviour will remain unaltered, in the sense that the proportion of each age class

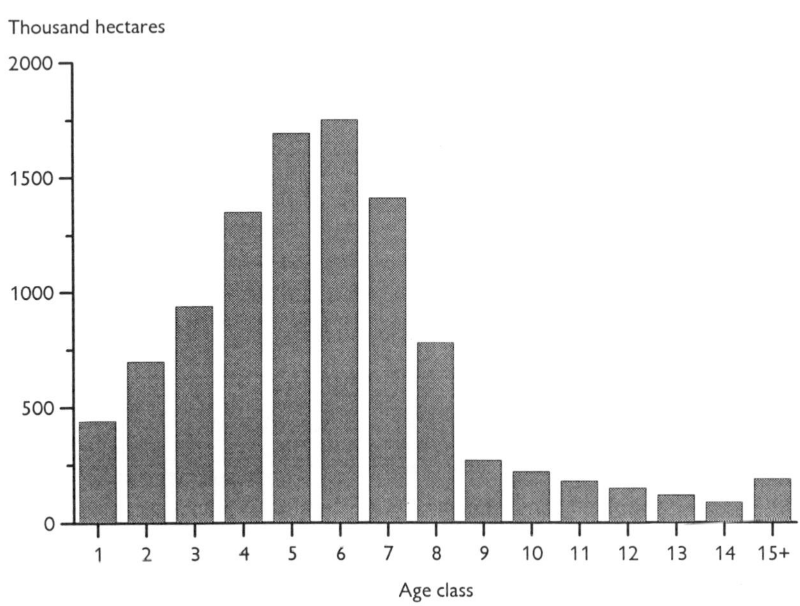

Fig. 1.7. The age class structure of man-made forests

felled remains the same, the output of coniferous timber should increase in proportion to the areas of forests in the age classes representing maturity. In this case, taking the age classes 8 to 15+ as being mature, the area will increase by about three or four times. This suggests that output will increase by a similar amount, to around 60 million cubic metres. This means that Japan would no longer need to import any coniferous timber.

These rough calculations show that the felling behaviour of Japanese forest owners could have a tremendous influence. The effects would extend not only to the domestic supply situation in Japan, but also to the foreign suppliers of coniferous timber. The west coast of North America, Russia's Far Eastern region, New Zealand, Chile and now even countries such as Sweden and Finland would feel the effects, as would competing importers such as Britain.

Natural forests show a slightly different age class structure and the pattern, using wider class intervals, is shown in Fig. 1.8. About 38% of the total area of natural forests is in age class 13+, compared with only 4% of man-made forests. However, there is a difference between national and private forests. Almost all of the nationally owned natural forests are in the higher age classes, 84% being in age class 13+. However, in the case of private forests the figure is only 14.5%. Private natural forests are more like man-made forests in this respect. Most old-growth broad-leaved forests are nationally owned while most of the second growth is in privately owned forests (Nishikawa, 1986).

A large part of the young, maturing natural forests in private hands is

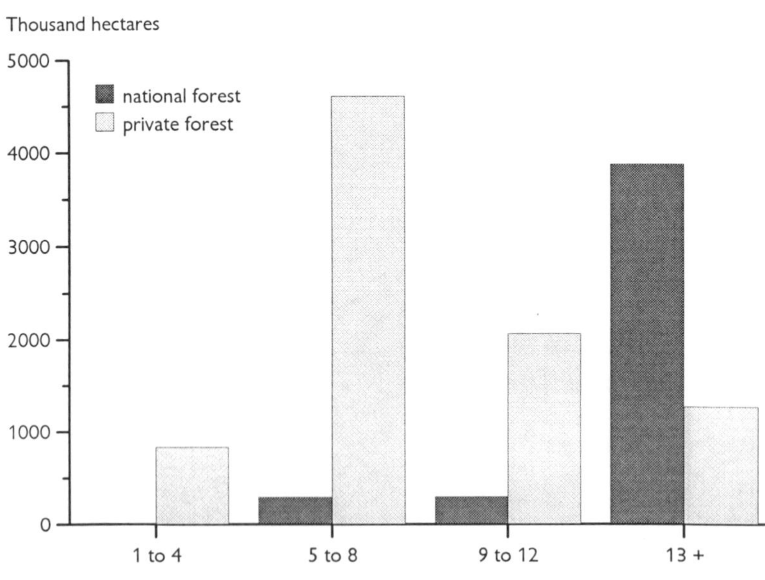

Fig. 1.8. The age class structure of natural forests

relatively low value broad-leaved trees which regenerated after the harvesting of firewood and timber for charcoal production in the years immediately after the Second World War *(Shinrin ringyou gyousei kenkyuukai*, 1988). Thus, the significant increase in supply which the age class structure of man-made forests could herald is unlikely to be reproduced in the case of natural forests. At best it means an increase in the supply of pulpwood and chips.

Concluding Remarks

The data described in Figs 1.7 and 1.8 underlie the discussion in the remainder of this book. As the simple calculations given earlier show, if the growing stock shown in the diagrams is actually felled the effect on world timber markets would be highly significant. On the other hand, the decline in the fortunes of the Japanese forest sector after the Second World War suggests that no such supply will be forthcoming. Indeed, apart from those intent on 'talking up the market', most commentators and most Japanese would agree that no major increase in domestic production is foreseeable.

However, it is not really satisfactory to extrapolate the decline in Japanese output into the future in this way. Any extrapolation of trends implies that the factors which have given rise to the data in the past will more-or-less continue into the future and Figs 1.7 and 1.8 show that this is not a reasonable assumption to make. The question posed, therefore, is will the decline in the economics of domestic timber production continue at a sufficient pace to choke off the increase in potential supply? In this sense, Japanese timber production is on a knife edge. The simple extrapolation of output represents a form of 'unstable' forecast. If the economics of forestry become more favourable the output of domestic timber could increase dramatically. If it becomes comparatively less favourable output could decline equally dramatically.

Against this backdrop of maturing timber and a declining rural economy, the countries supplying Japan with its timber are also experiencing problems of various sorts. The supply of tropical timber to Japan has well-known problems, Russian upheavals have meant that the output of timber products there has declined drastically and environmental concerns in North America are having an influence on their timber industries. On the other hand, countries such as New Zealand and Chile are beginning to experience an increase in output from their forests. Thus, the future potential to import timber into Japan is not something which can easily be predicted.

The remainder of this book examines these issues and culminates in the Japanese government's forecasts and the author's model of the timber market after discussing the situation in Japan and in the countries which supply Japan with timber. First, however, a more detailed look at the demand for timber in Japan will be undertaken.

Chapter two:

The Demand for Timber in Japan

Introduction

A more detailed view of the forest sector of Japan and its place in world markets is required if the objective of judging projections of future market conditions is to be fulfilled. As the long-term forecasts to be presented in the last two chapters are based on standard market models, it is at least consistent to break the discussion down into an examination, first, of the demand for timber and, secondly, its supply. In this chapter, the patterns of demand for various timber products and the factors that influence them will be examined.

Demand figures for timber products are regularly produced by the Ministry of Agriculture, Forestry and Fisheries (MAFF). The figures that MAFF typically publish are in the form of roundwood equivalents (RWE) with the factors used to convert, say, plywood panels into their roundwood equivalent varying slightly from year to year. The latest set of conversion factors are quoted on page *ix*. MAFF also publish imported timber and timber product volumes in unconverted units, these figures coming from the Ministry of Finance. Thus, the part of overall demand quoted in RWE and satisfied by Japanese production will correspond more-or-less to the harvested log volumes found in other MAFF data which will also be the figures used in Chapter three. However, the imported elements, if they are imported in processed form, will have been converted into RWE and so will not necessarily be identical to the figures for imports or exports quoted by other agencies.

MAFF divides timber demand into two broad categories; industrial use or *youzai* and 'others'. The latter contains firewood and logs for the production of *shiitake* mushrooms. The demand for these products will usually be satisfied in local markets and so the level of demand is not only small but also of no great significance for the international timber trade. The former category of *youzai* is further divided into *seizaiyouzai*, or saw logs and finished and semi-finished solid

Table 2.1. The demand for timber in Japan

	Demand (000 m³ RWE)	Domestic production (%)
Industrial wood (*youzai*)	112,324	20.0
Saw logs and lumber (*seizaiyouzai*)	49,758	32.5
Veneer logs and plywood	15,726	1.4
Pulplogs and chips	43,822	12.9
Other	3,018	14.2
Other uses		
Firewood	749	
Shiitake bed logs	967	
Total	114,040	21.2

source: Forestry White Paper (1997)

wood products, pulp and chips, veneer logs and 'others uses'. Other uses include transmission poles, pit props, timber piling and scaffold poles. The figures for 1996 are shown in Table 2.1.

As has been described in the previous chapter, the demand for timber in Japan increased quickly during the years of rapid economic growth, but since then has stabilized. The usual pattern of demand has been observed over these years with a decreasing proportion of timber going for saw logs and an increasing demand for pulpwood, chips and panel products and this is shown in Fig. 2.1. In the diagram, the category 'saw logs and lumber' includes 'other uses' shown in Table 2.1 as changes in definitions in the various sub-categories have distorted these two categories of product over time.

The per capita timber consumption, based on the figures for industrial wood in Table 2.1 and a population figure of 126.2 million, is 0.9 cubic metres. This compares with the British figure of around 0.78 cubic metres (FICGB, 1996). If the figures are broken down between end-uses an interesting difference between Japan and Britain emerges. The consumption of plywood (all panel products in the British case) was 0.11 cubic metres in both countries although FAO figures show a slightly higher figure for wood-based panel consumption in Japan than in the UK. However, FAO figures for 1996 show that, of the apparent consumption of wood-based panels, 69% was accounted for by plywood in Japan while the figure for the United Kingdom was only 23%. The per capita consumption of pulp and chips for 1996 was 0.36 in Japan whilst it was 0.37 cubic metres in Britain. The difference in the two countries' overall per capita consumption is almost entirely due to the higher level of consumption of saw timber and solid timber products. The British annually consume 0.30 cubic metres a person whilst the Japanese consume 0.40.

Differences in the conversion factors for calculating RWE mean that this comparison is at best only indicative but two peculiarities in the demand for timber in Japan emerge. One is the use of solid wood and the second is that the demand for panel products is very much dominated by plywood.

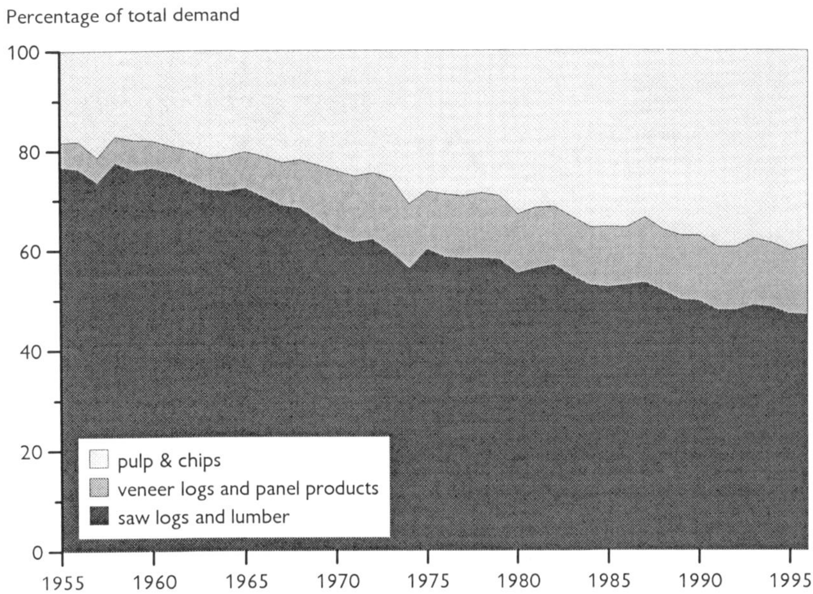

Fig. 2.1. Demand trends for the major product categories

In the following discussion each of these broad product categories will be looked at although, given the similarity in the end-uses of some of the products, a certain overlap in the discussion is inevitable. Because of the peculiarities shown by the per capita consumption figures the discussion will concentrate on sawnwood and panel products.

Saw Logs and Sawnwood

The demand for saw logs and sawn timber products is illustrated in Fig. 2.2. It exhibits the typical pattern for Japanese time series which has already been seen in Fig. 1.3. There is a rapid and almost constant growth until the first oil shock in 1973, followed by a slump and then the market picks up to give relatively stable demand. A final period of rapid growth during the years of the 'bubble economy' in the late 1980s is finally brought up to date with a relatively stagnant demand pattern. The per capita consumption figures grew in the early years but since then have been fairly steady showing, if anything, a slight downward trend.

Most Japanese authors argue that the main determinant of demand for saw logs and lumber in Japan is the demand for construction and especially housing. Some take what may seem a rather extreme view. For example, Yorimitsu (1992) states, 'it is no exaggeration to say that Japan's current wood demand is determined by the amount of domestic housing construction ... it is not too much

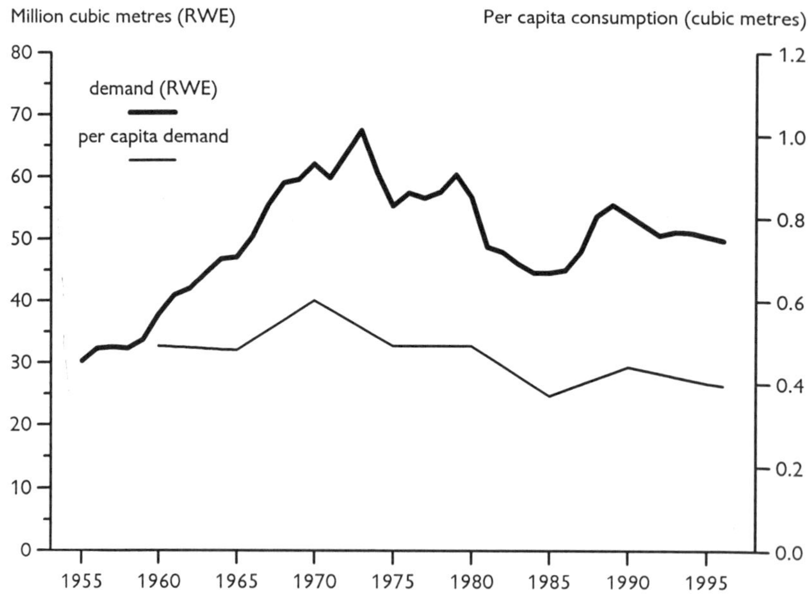

Fig. 2.2. The demand for saw logs

to say that the future of Japanese forestry depends on the trend of housing construction.'

The figures for the breakdown of saw log and lumber demand by end-use support this view. The 1996 Forestry White Paper shows that 71% of saw logs, sawnwood and solid timber products were used in housing construction (*jyuutaku kenchiku*), 9% in related building work with the remainder being used in civil engineering works (*doboku kenchiku*), for packaging (*konbouyou*) and in other uses. The figure for use in housing is likely to be greater for certain species, especially domestically produced sugi and hinoki. Because of their relative price premiums these trees are more prevalent in housing construction than the overall average figure of 71% would suggest.

This dominance of housing in the demand for timber has implications, not only for domestic forestry in Japan, but also for exporters. The housing market imposes special demands on suppliers of timber. Whilst Japanese suppliers are clearly more likely to be in tune with consumer demand, foreign exporters have to compete in this market and so an understanding of the demand patterns and trends is crucial. Also, it is in this market that the premium prices for timber can be obtained, unlike, say, the more conventional demand patterns seen for pulp wood or chips. It is no surprise to find, therefore, that housing demand is used as the main driver of timber demand in nearly all of the econometric models of the Japanese timber sector. So it is necessary to look at this market in some detail.

Housing demand

The supply of housing in Japan is regularly reported, either as the number of housing starts or as the floor area, and in 1997 the number of housing starts was just under 1.4 million units. This is a very high figure, comparable with that of the USA which has double the population and considerably in excess of Britain's, and this has been the case throughout the recent past.

Data for 1995 show that there were 2.82 members in the average Japanese household which, with a population of somewhat more than 120 million, gives around 44 million households. This figure implies that about 3.1% of Japanese households acquired new housing in 1995. If this is a stable situation it means that the average Japanese household will buy a new house once every 25 to 30 years, a figure roughly equal to the 30 years life-span for a house which the Japanese will usually quote.

The age distribution of the Japanese housing stock confirms this. Over 68% of the housing stock was built after 1970, compared to about 22% for Britain and 42% for the USA. Only 7.3% of houses were built before 1950 compared with about half of the housing stock in Britain (*Mokuzai jouhou*, 12/96).

Thus, the very high level of housing starts is mainly a reflection of the 'scrap and rebuild' approach to housing found in Japan, which, together with the fact that almost half of the houses built are wooden, creates a very high demand for lumber. This attitude to housing, some commentators argue, is a reflection of the inherent insecurity that the Japanese feel in the face of the elements. The fact that Japan is situated in an area with frequent earth tremors and feels the effects of the 'typhoon' seasons and flooding referred to briefly in the previous chapter leads to the idea that housing is not an item to which the characteristic of permanence may be attributed. Similarly, it is true that the DIY booms in Europe and America have not really been reproduced yet in Japan and the average Japanese householder does not spend anything like as much time in maintaining the house as would be the case elsewhere.

The short life of Japanese houses can lead to an instability in the level of demand in that market. The decision to scrap and rebuild is one which can, in most cases, be fairly easily postponed for a year or two. Thus, an economic downturn can see a rapid fall-off in demand as rebuilding plans are set back. This may well be the case in the 1997 figures for housing starts which represent a drop of 14% on the previous year's level.

Figures for the number of housing starts (*shinsetsu jyuutaku chakkou kosuu*) and the average floor area are shown in Fig. 2.3. The diagram shows an almost identical pattern to that exhibited by the total demand for saw logs in Fig. 2.2. Until 1973 housing starts grew steadily. In the early years after the Second World War, on average, 266,000 housing starts were recorded each year and in the era of high economic growth this figure rose to 915,000. At the same time the economy was growing by between 9 and 10%. In 1973 two things conspired to halt that trend. One was the oil shock and the other was the fact that, in that year,

Fig. 2.3. Housing starts and average floor area

the number of houses finally surpassed the number of households (Yoshida, 1981; Wakimoto, 1998).

Since that time the level of housing starts has remained more-or-less stable, showing neither a significant upward or downward trend whilst real GDP growth had slowed considerably. However, the increasing quality of the housing, especially when compared with the low quality 'barracks' which were produced after the war as a short-term expedient, probably means that the life of the recently built housing will be somewhat greater. So the stable level of housing starts is consistent with a slowly growing housing stock.

Figures show that, like other countries, the average size of the household is declining. Whilst the trends are not as advanced as they are elsewhere, the level of divorce and the increasing number of nuclear instead of extended family units have driven the increase in the demand for housing. In 1950, for example, the population of Japan was around 84 million and the number of households stood at 16.5 million, meaning that the average household size was just over five persons. By 1990 this figure had dropped to just three and in 1995 it had fallen further to 2.82. With the increase in population this has led to the number of households expanding to just over 43.9 million, an average annual increase of around 2%.

The increasing 'westernization' of the country would tend to suggest that these trends will continue and family sizes will become smaller. Recent research published by the *Jinkou mondai kenkyuushou* (Population Problems Research

Centre) and reported in *Mokuzai jouhou* (8/98) suggests that, although population is expected to peak in 2007, the number of households will to continue to increase. Their projections are shown in Fig. 2.4.

The total number of households is projected to rise from the 1995 figure of 43.9 million to 50.2 million in 2010. This is an annual growth rate of a little under 1%. However, the number of nuclear families is expected to increase more rapidly, along with the number of households comprising only individuals. The growth rates of these two groups are 1.3% and 1.6% respectively. The only group expected to show a decline is the category 'others' which contains extended families. The research further divides the nuclear families into those with children and those consisting of just couples. The former, whilst being the larger, is expected to grow at only a little more than 0.5% over the period whilst the childless 'families' are projected to show a growth rate of over 2.5%.

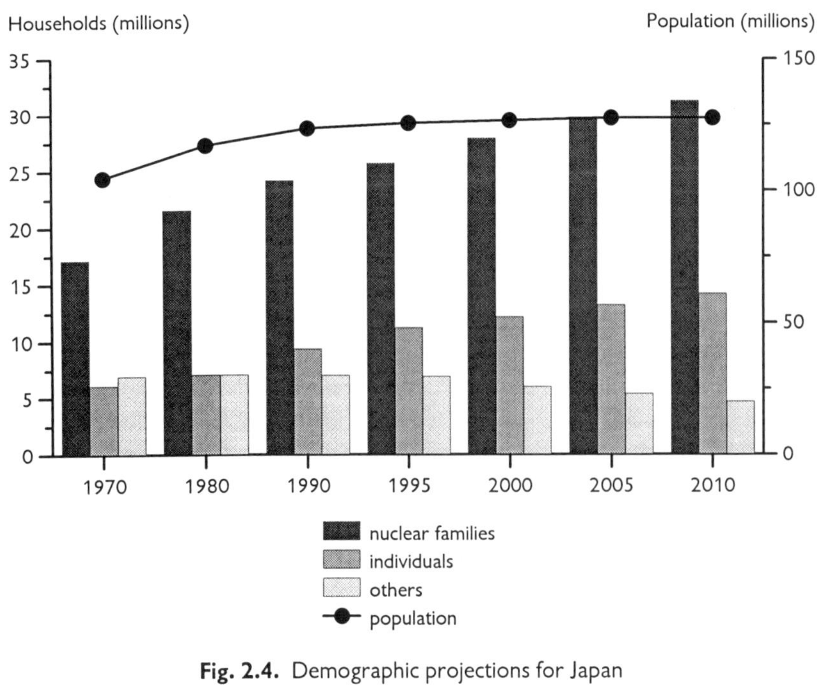

Fig. 2.4. Demographic projections for Japan

The forecasts for housing starts which accompanied these projections of demographic change suggest an increase in the housing stock but a decrease in the annual level of housing starts. These forecasts are shown in Table 2.2 where the total number of households in the second column corresponds to the sum of the various categories of household shown in Fig. 2.4.

The required housing stock in the forecasts is simply taken as a multiple of the number of households, the factor being 0.962. The actual housing stock seems to be calculated on the basis of a fairly static vacancy rate of around 12 to 13%. The annual 'new' housing starts are simply the difference in the actual housing stock divided by five. The 'replacement' housing starts are calculated on the basis of the age profile of the existing housing stock and this, with the new housing starts, gives the annual level in the final column. The figures show that the increase in the housing stock implied by the demographic forecasts in Fig. 2.4 translates into a falling level of housing starts.

Table 2.2. Forecasts of housing starts (thousands of units)

	Total number of households	Required housing stock	Actual housing stock	New housing starts	Replacement housing starts	Total housing starts
1995	43,521	41,867	47,184	649	773	1,422
2000	46,145	44,391	50,162	596	852	1,448
2005	48,371	46,533	52,582	484	905	1,389
2010	50,180	48,274	54,550	394	949	1,343

source: *Mokuzai jouhou* (2/97)

The table is interesting in dividing the housing start figures into new housing and replacement demand. The latter is rising and, it could be argued, it will actually rise by more than the figures here suggest. The breakdown of the age of the housing stock referred to earlier on page 31 showed that, while 6.5 million housing units were built between 1961 and 1971, there were 12.5 million housing units which had been built between 1971 and 1980 and a similar number in the subsequent decade. This suggests a rapidly increasing replacement demand during the period of the forecasts in Table 2.2, a trend which the Japanese forecasts do not seem to capture. It would seem permissible, therefore, to argue that the prognoses in Table 2.2 are at the pessimistic end of the spectrum and that the future trends in housing starts could well be stable or increasing rather than declining.

The trends in the different types of households shown in Fig. 2.4 clearly have implications not only for the level of housing demand but also for its composition. The high cost of land and the centralized nature of the economy causing population concentrations in the big urban areas mean that the ability of individuals or small families to purchase their own homes is restricted, suggesting a moderating influence on increases in the average floor area. The structural shifts suggested by the research would also be compatible with increases in the number of households living in apartments rather than individual houses and this, as will be seen, is likely to mean a decrease in the demand for lumber but an increase in the demand for plywood and other panel products.

However, demand for timber depends on the area of the housing starts as well as just the number of starts themselves. Figures for housing starts expressed in terms of floor area follow a very similar pattern to that in Fig. 2.3. However, as

Fig. 2.3 showed, the average floor area rose in the period before the oil shock. After that it fell but rose again during the years of the so-called 'bubble economy'. Thus, the total floor area showed a somewhat more exaggerated rise and fall than that shown by housing units. International comparisons of floor area are a little difficult to make as the Japanese define floor area somewhat differently and the areas which are included as 'space' differ. But, in general, newly built houses in Japan are not much smaller than those in Europe although they are smaller than those in the USA. One difference is the proportion of the plot which is not built upon. This seems to be considerably lower in Japan than in other countries, giving Japanese housing and towns a cramped feel.

A factor that could act to increase the average floor area is the recent legislation allowing three-storey wooden houses to be built. Until recently, two storeys was the maximum height for wooden housing, with the obvious exception of temples. However, partly under American pressure and partly because of the lack of space in the major Japanese cities, it is now permissible to build three-storey houses from wood. Surveys suggest that this form of housing could be popular with respondents stating that they will be more airy and that in winter the rooms will be more sunny. In reality, however, it is likely that the only view from the third-floor will be the third-floor of the neighbouring three-storey house.

Also, there has been an increase in two or three generation housing (*ni/san sedai juutaku*). It is not uncommon for housing to be built in which two or three generations live. Often the interior of the houses are separated so that a single house becomes, effectively, two or more apartments for members of the same family. However, Fig. 2.4 suggests that the number of such households is set to drop in the future.

Overall, then, it is difficult to make projections about the average floor area. With a growing economy trends which have already been observed in other countries can be projected onto Japan with homes getting larger and children having their own rooms. On the other hand the demographic projections and pressures of space suggest a possible reduction in the average figure.

There are some trends within the data for housing starts which are of significance for the analysis of timber demand and for those hoping to supply timber to the Japanese market. These hinge on the ways in which housing might be classified and there are a number of sub-categories into which the Japanese divide their housing statistics. Important for the discussion here is the division between 'wooden' (*mokuzou*) and 'non-wooden' (*himokuzou*) housing and the division by end-user. These will now be looked at.

Wooden and non-wooden housing

The distinction between these two types of housing is based on the material used for the main framing of the house. If timber is used for the pillars, beams, staircases and sills on the concrete foundations the building is classified as 'wooden'. The significance of the wooden housing sector can be ascertained from figures from the Ministry of Construction. According to them, in 1993 the demand for timber for wooden housing accounted for 77.4% of the total amount of timber going into construction. Figures from MAFF on page 30 showed that 71% of sawnwood was used in construction. This implies that about 55% of all lumber demand was accounted for by wooden housing.

The division of buildings into wooden and non-wooden gives rise to the *mokuzouritsu* or proportion of wooden housing, quoted either on the basis of housing starts or floor area. The overall figure for this ratio has been falling whether the figures are based on floor area or housing starts, and both are shown in Fig. 2.5.

The proportion has dropped more quickly if the number of housing starts are used to calculate the ratio. This reflects the fact that the average floor area of wooden houses is greater than that of non-wooden ones. For example, in 1997 the average floor area of newly built wooden houses was 112 square metres compared with only 78 for non-wooden ones. If the figures for 1965 are examined, the average floor area for both of these categories was only slightly less than 60 square metres, so wooden housing has shown a significantly greater improvement over the years than has non-wooden.

The alternatives to wooden housing, especially the concrete-based apartment blocks known, apparently without irony, as 'mansions' have become increasingly popular. This is largely due to the high land prices not allowing younger couples to purchase individual houses which are more likely to be wooden. However,

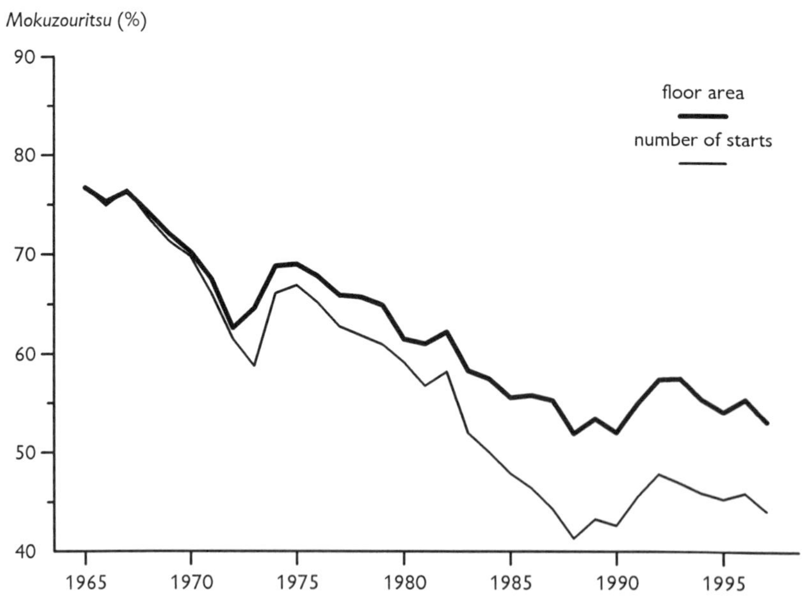

Fig 2.5. The proportion of wooden housing

while Fig. 2.5 shows a long-term downward trend, the *mokuzouritsu* also rises in the early 1990s. This suggests that when GDP rises 'abnormally' wooden housing is preferred implying that non-wooden housing is an inferior product in the economic sense. This is something which the Japanese would tend to confirm. Surveys have continually shown that only around 14% of the population would actually choose non-wooden housing in preference to wooden.

Not only do broad macroeconomic and demographic trends affect the choice of housing but local differences also exist. There is, for example, a correlation between the forest cover of a prefecture and the ratio of wooden housing, the correlation coefficient being 0.50 which is significant at the 1% level. This high figure is probably indicative of the economies found from local markets for domestically produced timber. It would also suggest that, if the local markets have been and still are of importance in the demand for timber, there are likely to be local differences in the type of timber used and the characteristics of that timber that are valued - some of these are mentioned below (*Mokuzai jouhou*, 11/96).

It is not straightforward however to draw a direct linkage between wooden and non-wooden housing and the demand for timber. On the one hand, the lower the proportion of wooden housing the higher is likely to be the demand for plywood as this will be used as concrete shuttering in the construction of non-wooden housing. On the other hand, apartments in concrete blocks will contain traditionally styled rooms reminiscent of wooden houses in the Japanese style. Conversely, individual houses are sometimes constructed of concrete. There is an important overlapping division of housing which affects the demand for timber, that is between 'traditional' or 'Japanese styles' and 'non-traditional' or 'western styles' of housing.

Traditional and non-traditional styles

The term 'traditional' is not one which is well-defined for the purposes of statistics. In general, the traditional, Japanese-style house will be built with a post-and-beam method where vertical pillars are joined to horizontal beams. This means that the walls themselves are relatively unimportant in the structural sense. The spaces between the pillars can be fitted with sliding doors allowing the effective sizes of the rooms to be altered. This aspect of design is, in part, a response to the hot, humid summers and cold winters described in Chapter one. Traditional Japanese houses have tatami or rush matting floors rather than wooden or carpeted ones. In Japanese-style housing the wall studs have smaller widths than the columns allowing the wall thickness to be less than that of the column. This leaves part of the column visible. In non-traditional or western construction the walls would be such that the columns are concealed, leaving the usual flat wall to be decorated with paper.

Hinoki and sugi are the preferred species in most cases for columns with hinoki being the more expensive. However, sugi from natural stands or from forests with a long tradition of supplying timber such as those where serious production forestry began in the Tokugawa era, is considered highly desirable and commands a high price. Ezomatsu from natural stands in Hokkaido or imported from Russia is also a popular species for columns, especially in Hokkaido. Wood imported from America is commonly used for inexpensive columns. Akamatsu is used for beams that are subjected to large shear stresses. Hinoki, karamatsu and timber from America are used for sills, with chemically treated imports gaining ground. For beams and floor joists akamatsu and karamatsu are popular with American imports also being much used. For components when length is of importance, where solid wood was once used, now plywood panels or laminated timber are taking over.

Having made these points it is true to say that within Japan, as in all other countries, there are various 'traditional' styles. These often relate the varied climatic conditions found in the country and briefly outlined in Chapter one. For example, the high levels of snow in Hokkaido and the northern and eastern parts of Honshu mean that stronger housing is called for and so column sizes are typically larger. There are now two sizes for columns in use in the country. One, a 12 cm square section, is used for greater strength whilst the other, 10.5 cm square, is common in the southern parts. In the more humid and warmer southern parts of the country greater emphasis is placed on sliding panels to allow a through-flow of air whilst in Hokkaido warmth is more important.

As well as these climatically determined variations, tradition plays a part. As a general rule, redder timbers have been more popular in the north of Japan and so the white wood of hinoki is less popular there than in the south. Akamatsu is a popular building material in the Kanto area and on Shikoku. Also, given the relatively recent population movement to Hokkaido, there is less of a traditional form of house building and so non-traditional housing is dominant there, with price being a major determinant in the choice of material.

Clearly, therefore, the proportion of traditional housing is an important determinant of timber demand. Unfortunately, however, there is no easy way to measure this, indeed it is usual for styles to be mixed within one house. One set of figures that is published is the number of *washitsu* or Japanese-style rooms in a house. In 1992 the average number of *washitsu* was two per house or apartment. This figure was down from the average of 2.4 in 1988 (*Mokuzai jouhou*, 11/96). However, despite this decline it appears that the number of houses without a *washitsu* is only around 3% (*Mokuzai jouhou*, 3/97). Another variable of relevance is the so-called *hikidoritsu* which shows the proportion of housing which has a sliding door at the main entrance. This, again, is a Japanese tradition and western-style housing will have ordinary hinged doors. Although there is an increasing trend to use aluminium for such sliding doors their use indicates a more traditional style of building and, therefore, a probable greater use of visible timber. The 1992 figure for this proportion was 37.9%.

Another figure of significance is the number of houses in which there are two adjoining *washitsu*. This might seem a strange statistic but is one of the features of the Japanese style outlined above with adjoining *washitsu* allowing flexibility in room layout. As would be expected, in prefectures where the number of *washitsu* is relatively high the number of contiguous Japanese-style rooms is also high (*Mokuzai jouhou*, 3/97). The figure for Hokkaido is only 15.3%, the lowest in the country, and it is around 20% for the large cities areas. It is here that the falling number of *washitsu* might become critical. If only one *washitsu* is contained in a house it is obvious that one of the features of Japanese architecture will be impossible to realize and, so, the Japanese-style room will be of ornamental value only. The question would, then, become one of whether the Japanese would continue to be willing to pay for something which is of purely cosmetic value.

The pressure of space and the perceived extra convenience of western-style housing, as well as cost considerations, is why there is a trend in the market away from traditional housing towards western styles. But it is in the Japanese styles that the premium prices for timber can be obtained and where exporters are at the greatest disadvantage. In western-style housing it is often the exporters of timber who can exploit their cost advantage and so drive Japanese timber out of the market. Indeed, it is usual to see partially constructed houses with knotty pillars in place in areas where their surface will not be visible when the house is finished and the pillars destined for *washitsu* carefully wrapped to prevent damage to their surface.

Firmly in the non-traditional type of housing are so-called 'prefab' and 'two-by-four' construction methods. The prefab method, where construction components are prefabricated in factories to reduce on-site construction costs, appeared in the late 1950s when the rate of housing construction was rapidly increasing. The two-by-four system was developed in America. Structural plywood is nailed onto wood components which are assembled to make the finished house. This method was allowed as a construction method in 1974, but in 1982 it was claimed that the method 'is relatively new, it still is taking time to develop a system of material acquisition and to train technical personnel. Therefore, the number of houses built by this method is still insignificant.' (JFTA, 1982). However, since then this housing method has been growing in popularity. Figure 2.6 shows the recent trends in these types of housing construction. Although post-and-beam maintains its dominant position in the market, the two-by-four method is rapidly gaining market share.

The preference for post-and-beam construction, or at least the inertia of those who produce it, is reflected in the more modern three-storey wooden housing with around half being constructed on this basis. The proportion of two-by-four housing in the three-storey market is about 18%. The remainder is made up of prefab type construction and mixed construction where a steel ground floor supports two wooden storeys, these often being built in the post-and-beam method. The preponderance of two-by-four housing in this new category is slightly greater than in wooden housing as a whole (*Mokuzai jouhou*, 9/96).

Surveys of housing preferences have long been conducted and the preference

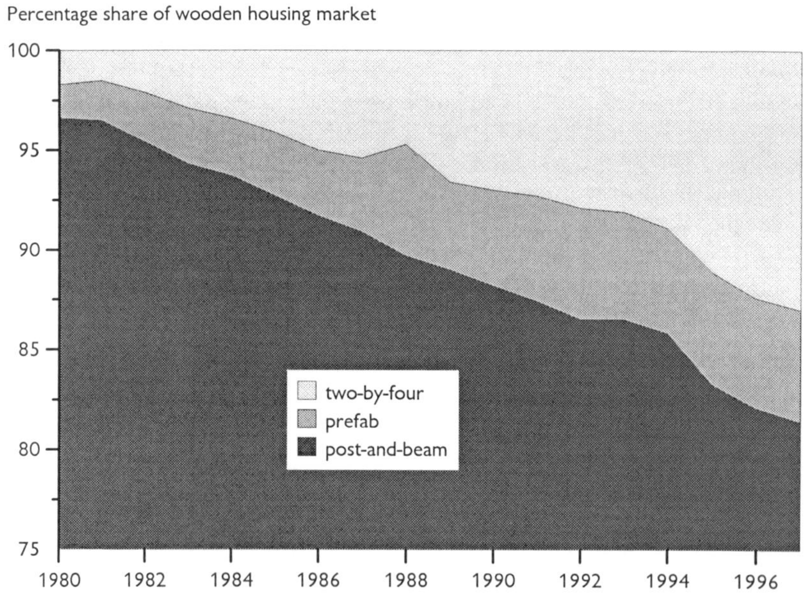

Fig. 2.6. Two-by-four, prefab and post-and-beam housing

expressed for two-by-four housing is increasing. In 1976, for example, only 4.7% of those questioned answered that they wanted to live in prefab or two-by-four style housing. The largest response, 74.8%, was not surprisingly for post-and-beam housing *(Shinrin Ringyou Gyousei Kenkyuukai*, 1988). The latest survey in 1996 still puts post-and-beam construction as the most popular with 69.1% of the sample saying they want to live in such a house. But the percentage wanting to live in prefab or two-by-four type housing has risen to 12.5%. The category of 'non-wooden' housing has attracted a relatively constant 14 to 15% over the years (*Rinya jihou*, 9/96).

The construction of two-by-four style homes has opened the door to house as well as just timber imports. At present the overall level of such imports is relatively small with most of the imports being of the two-by-four type (Hashimoto, 1995). A feature of the Japanese market, which is gradually being eroded with the use of more factory-based housing production systems, is the high level of houses that are built for purchasers on an individual basis. This favours the small builder working on single contracts and tends, therefore, to create a favourable environment for the post-and-beam construction method.

To compete against the increasing inroads being made by two-by-four and prefab techniques so-called 'precut' (*purekatto*) methods are being used. In their idealized form, CAD/CAM systems are used to design houses to fit individual needs and the components required are cut and delivered for assembly on site. At the beginning of 1997 there were reportedly 865 factories operating 977 lines in this field, most of which are new, the boom in factory establishment being after 1990 (*Mokuzai jouhou*, 7/95, Takaki, 1997). The method has proved popular with it now reportedly accounting for around a third of the market for wooden housing, with the figure being as high as one half in the urban centres (Takaki, 1997). Although the basic idea seems to have been to improve the efficiency of the traditional housing sector, the wood used by these factories is predominantly imported from the USA. The proportion of domestic timber used by 'precut' producers in 1994 was 38.1%, only slightly above the overall self-sufficiency ratio for lumber as a whole (*Mokuzai jouhou*, 7/95).

The average amount of timber used in wooden housing is given in Table 2.3. Whilst the overall figures for timber consumption for two-by-four housing are similar to those in traditional post-and-beam methods, the nature of the timber used differs to the extent that there is less visible timber in the newer construction methods. Therefore, the ability of lower quality imported timber to compete in Japanese markets will be increased with the increasing use of two-by-four housing methods.

Table 2.3. The use of wood in wooden housing

	Floor	Wall	Roofing	Ceiling	Other	Total
Post-and-beam	0.057	0.065	0.046	0.008	0.014	0.191
Two-by-four	0.068	0.094	0.046	0.004	0.008	0.221
Prefab	0.043	0.054	0.025	0.015	0.017	0.153
	Structural	Furring	Fixtures	Finishing		
Post-and-beam	0.138	0.028	0.022	0.003		0.191
Two-by-four	0.200	0.005	0.014	0.002		0.221
Prefab	0.095	0.045	0.012	0.001		0.153

All figures in cubic metres per square metre of floor, total figures contain rounding errors.
source: *Mokuzai jouhou* (3/95)

Non-wooden housing (*himokuzou juutaku*) takes a number of forms and, of course, uses a certain amount of wood in its construction. As emphasized earlier, the fact that it is non-wooden does not imply that it has no traditional elements. The use of timber in non-wooden housing is shown in Table 2.4 and it can be seen that the figures are significantly below those registered in wooden housing of any sort.

Table 2.4. The use of wood in non-wooden housing

	Ferro-steel	Reinforced concrete	Steel frame
Residential housing	0.040	0.051	0.048
Mixed housing & commercial	0.037	0.048	0.041
Offices	0.019	0.034	0.012
Industrial premises	0.018	0.022	0.008
Educational establishments	0.030	0.034	0.025

All figures in cubic metres per square metre of floor.
source: *Mokuzai jouhou* (3/95)

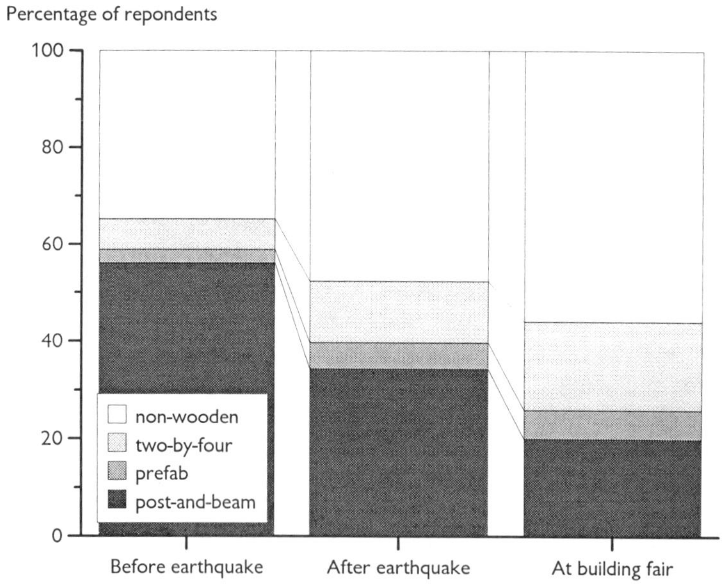

Fig. 2.7. Housing preferences and the Kobe earthquake

One of the features of the Japanese housing market is the often stated position that Japan has to have wooden housing because of the climate and the frequent earthquakes felt in most of the country. The Japanese make the point that timber is good at absorbing moisture and so is useful in reducing the humidity of houses in the summer months, tending to ignore the implications for dimensional stability and decay. However, it would seem that the Japanese themselves do not really believe this position. Three surveys of housing aspirations taken before and after the Kobe earthquake of 1995 showed an interesting change in opinions. The answers to the question 'what type of housing do you prefer' are shown in Fig. 2.7 (Kata, 1995; Murashima, 1996).

The figures before the earthquake show a somewhat lower preference for wooden housing than that usually recorded in national surveys and reported above. However, the change in attitudes after the earthquake is marked. The third column in Fig. 2.7 shows the result of a survey of 813 people at a building fair held in Kobe after the earthquake. Here, the difference is even more marked with the preference for non-wooden housing surprisingly high and traditional housing showing a very low result indeed. The commentary with these results suggested that part of the reason for this was the fact that many of the houses which collapsed in the earthquake were badly built, old wooden ones, constructed at a time when regulations exhibited greater laxity. However, whilst many Japanese specialists suggest that the relatively poor performance of post-and-beam houses in the earthquake is the result of their age and that building regulations have since been tightened, the view from abroad appears to be that the problems are inherent in the post-and-beam style itself (New Zealand Pine International, 9/95).

The figures need a little care in their interpretation as the Kobe area has an above average proportion of non-wooden housing. Figures for the year before the earthquake show that only 35.2% of housing starts were in the wooden housing category compared with the national average of 46% - see Fig. 2.5 (Kata, 1995). However, if the trend shown in Fig. 2.7 is typical the effects would be profound. The figures in Table 2.4 show that the demand for timber would fall very rapidly and most of the Japanese growing stock would no longer be able to command a price premium with the bulk of the demand for timber being for plywood or non-visible structural timber. Indeed, short-term forecasts for the country as a whole suggest an increase in the amount of prefab and two-by-four housing at the expense of post-and-beam (*Mokuzai jouhou*, 2/98).

Classification by end-use

Figures for traditional and non-traditional housing starts are not available, and, as the discussion above suggests, they would make little sense. The figures which are easily available break housing down into another grouping. The four categories which are usually reported are *mochi-ie*, *bunjou*, *kashi-ie*, and *kyuu-you*.

In general, *mochi-ie* housing will be constructed for a specific owner who intends to live in the finished house. The houses in this category are typically purchased by individuals and built for them on land which they own. The houses are usually constructed by local builders and would be individually designed. These houses are far more likely to be of the wooden type. Indeed, the *mokuzouritsu* for *mochi-ie* is typically now around 75%. The data show that this ratio has fallen, like the overall figure, from 89.9% in 1965 to 73.8% in 1995, using figures based on floor area. It is in this category that the traditional Japanese-style housing is mostly to be found and post-and-beam methods would be the most common. It is usual for the purchaser to visit the house during its construction and assess progress. This means that builders tend to use material which a non-specialist would judge as high quality. This favours the consumption of well-known species and knot-free timber.

The term *bunjou* literally means to sell in lots and it usually implies that houses have been constructed on relatively large plots of land and then sold individually. *Bunjou* housing is built with no specific purchaser in mind and is purchased during or at the end of its construction. This type of housing is typically the preserve of the large companies, partly because of the capital that is tied up in the purchase of the site. This type of construction has a relatively poorer reputation in terms of housing quality than does the *mochi-ie* category because, as a general rule, the potential owners inspect the property and purchase after completion rather than visit the site during construction. This means that the purchasers do not see the timber which goes to make up the frame of the house, or indeed the walling. The suspicion is, therefore, that lower quality timber is used in this form of construction than in others. It is in this category that a lot of the prefabricated methods are used, and specifically two-by-four housing.

The *bunjou* type of housing includes mansions and therefore contains a large proportion of concrete construction. Recent figures suggest that about 57% of *bunjou* housing are mansions. The *mokuzouritsu* for *bunjou* housing is, therefore, a lot lower than that of *mochi-ie*, standing at the moment at around 43%.

Kashi-ie housing is built specifically to be rented by its owner to others. As with most rented accommodation, the quality of this housing is likely to be poorer than that of the other categories. The size of the housing development will vary. In many cases a small house or a block of four to six apartments in a two-storey wooden building will be constructed on land that the owner has no other use for. On the other hand, the construction could be a large mansion-style block owned by a large company specifically serving the rental market. If it is of wooden construction, and the present proportion is only just over 18%, the timber will be of lower quality to that found in *mochi-ie*, often being knotty. However, as the low proportion of wooden housing figure shows, a large part of this category is likely to be of concrete construction.

Finally, *kyuu-you* housing is owned by companies and is intended for use by their employees. It is usually non-wooden housing, being often of mansion style. The present the *mokuzouritsu* is just over 16%. In recent years companies have tended to subsidize the purchase of accommodation by their employees rather than build it themselves and this, to some extent, explains the reduction of this form of housing.

The relative proportions of these housing categories are shown in Fig. 2.8. The two most important categories are *mochi-ie* and *bunjou*, with the latter increasing over recent years.

The average floor area of the various types of housing starts differ and this will also impinge on timber demand. Figures for 1996 show that, against an average floor area of 96.1 square metres for all buildings and 116 for wooden housing, the largest houses were in the *mochi-ie* category with an average floor area of around 140 square metres. Next came the *bunjou* type of housing with average floor areas of between 100 and 116 square metres. As would be expected, the lower quality housing for rent had a floor area of only around 50 square metres. Thus, the increase in *bunjou* housing shown in Fig 2.8, if continued, would tend to reduce overall average floor areas slightly and therefore have a dampening effect on timber demand.

Concluding remarks

The discussion above is cold comfort for Japanese forest owners. Housing styles are moving away from the types which favour high-price, high-quality timber where Japanese forestry might be able to compete and moving towards lower cost housing where the visual characteristics of timber are of less importance. The

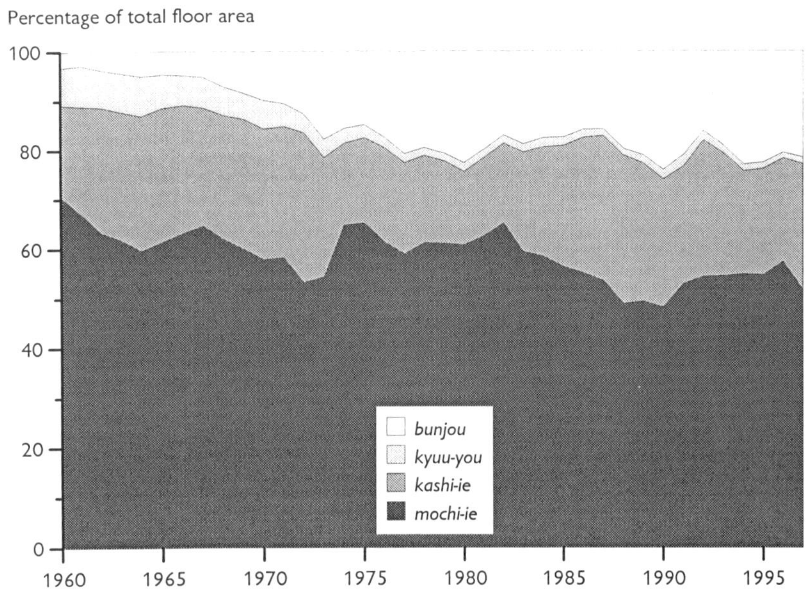

Fig. 2.8. Types of housing start

traditional Japanese house, with exposed timber, requires a quantity of good quality timber, often of specific traditional species. Newer western types of construction, however, have far less exposed timber. The demand for timber is, therefore, becoming more 'conventional' in nature, allowing greater competition on mechanical characteristics. This will enhance the comparative advantage of foreign suppliers, especially the North American lumber producers.

As will be seen in the next chapter the output of sugi has been holding stable in recent years against a decline in overall domestic timber output despite these trends. However, the reasons do not seem to be because of any inherent 'Japanese-ness' of the timber. Rather a recent survey of timber users found two main causes. One was that the price differential between sugi and American timber had narrowed suggesting competition purely on price. The second cause was that the proportion of *washitsu* in houses was decreasing as was the number of post-and-beam houses implying that sugi's visual characteristics are now of low importance (Suzuki, 1997). Apart from the ability of sugi to compete, the demand for hinoki in good quality, post-and-beam housing with large numbers of traditional rooms is falling as is the price premium it can command over other species.

The only trend that might mitigate against these movements in the housing market is so-called 'precut' housing. In other areas of industry Japan has proved capable of competing in terms of technological innovation and the manufacture of products attuned to consumer demand, but the ability to reproduce this success in the housing market seems limited. The large companies producing *bunjou* housing will probably always be better placed to compete on cost in the housing market.

Using the figures in Table 2.3 and Table 2.4 it is possible to calculate a form of 'elasticity of demand' for timber in housing. If the present *mokuzouritsu* were to drop by 1% then, using the levels of housing starts and floor areas referred to earlier in the text, the demand for timber for housing would fall by 1.5%. Thus, a continued move to mansion-style accommodation instead of *mochi-ie* type housing would lead to a proportionally greater drop in timber demand. Of course, within the wooden housing category, a shift from post-and-beam methods to two-by-four houses would have a relatively small effect on timber demand as Table 2.3 shows. But it would increase the demand for panel products.

As the movement towards mass-produced housing units will increase the use of panel products in housing and it is now time to examine the demand for these products.

Plywood and Veneer Log Demand

Unlike most countries, where the demand for panel products encompasses a wide range of products, in Japan by far the most dominant panel product is plywood. Although the figure has been falling in recent years, in 1994 plywood still accounted for three-quarters of the panel product market in Japan and the Forestry White Papers still do not have separate categories in their main statistical annexes for other panel products such as medium density fibreboard or particle board.

Plywood demand in Japan has also been an area of controversy because of its association with tropical hardwood log imports. Until fairly recently, by far the major proportion of plywood demand in Japan was satisfied by timber imported from Southeast Asia and this involvement with the tropical timber trade is a cause of friction between Japan and outside environmental organizations.

The plywood industry was different from the other areas of the Japanese forestry sector in that it was a sector which exhibited, in the early days, a flourishing export trade to the United States. The industry was favoured by government policy in the early post-war years as a means of earning hard currency. Foreign exchange was released for the purchase of logs from Southeast Asia with the intention of producing plywood for export to North America. The result was that both production and exports rose rapidly, along with the imports of raw material from Southeast Asia.

This trade was initially stimulated by the demand for flush doors in North America as part of the housing boom there after the Second World War. Tropical hardwood plywood had the characteristics to fit this market. It was cheap and had a plain appearance which allowed it to be easily stained, painted or printed. Originally the Philippines supplied the United States as the plywood industry in

Japan had largely been destroyed in the war. But as the industry re-established itself it was able to supplant the smaller, less efficient Philippine producers and become the dominant supplier to the United States using timber which, ironically, was mostly imported from the Philippines. At its peak in the mid-1950s about one-third of output was exported.

However, increasing domestic demand, coupled with increasingly export-oriented plywood industries in Korea and Taiwan, meant that the amount of Japanese plywood going for export began to diminish and, by the 1970s, exports had become an insignificant proportion of output. By this time the plywood industry had become, like the rest of the Japanese forest products industry, essentially inward-looking for its markets and using predominantly imported raw materials (Nomura, 1981; Cengel and McKillop, 1990).

These trends are shown in Fig. 2.9 which plots the production and trade of ordinary plywood. The apparent consumption of plywood, defined in the usual way as production plus imports minus exports, shows the typical pattern of a steeply rising trend until the oil shock of 1973 followed by a fairly stable level of demand. Now the self-sufficiency ratio in the plywood sector, defined in RWE, is only 1.4% which is the lowest of the main categories of timber demand (Forestry White Paper, 1997).

Recently there has been an increased import of plywood and the concomitant drop in domestic production based on imported logs. The low self-sufficiency ratio, of course, results from the fact that almost all domestic production is based

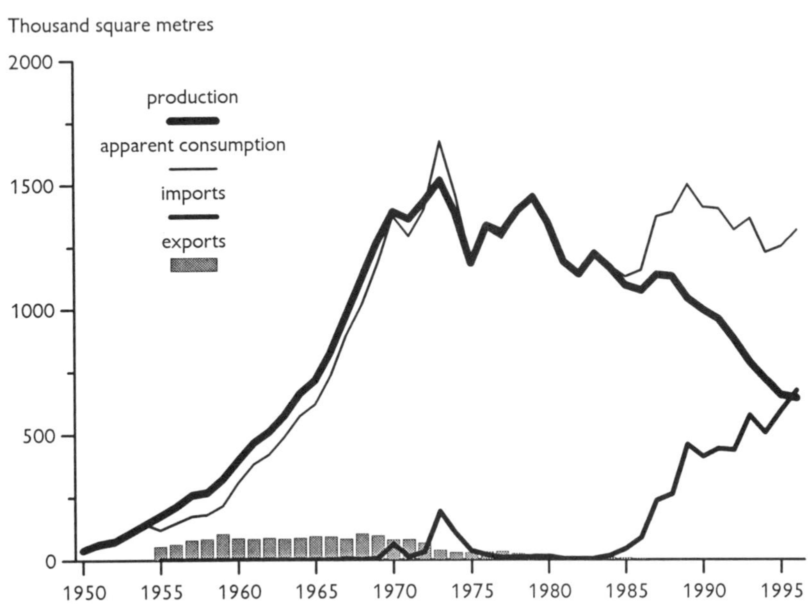

Fig. 2.9. Plywood production and trade

on log imports. The reasons for this reduction in domestic production are largely the growing plywood industries, often backed by log export bans, in Southeast Asian countries and the increasing price of logs.

At present, over half of the plywood consumed in Japan is used in construction and the second biggest user is the furniture industry. The use in housing is dependent less on the traditional types of housing but more on the modern methods of two-by-four or concrete-based apartment blocks. However, the demand from the construction industry comes not just from domestic housing but also from general building where the material is used as shuttering and it is here that the controversial use of tropical hardwood plywood for concrete moulding (*kon-pane*) is found.

The increasing costs and scarcity of tropical logs and, recently, environmental pressures to reduce imports of Southeast Asian hardwoods, have led the Japanese to attempt first to diversify their sources of supply and secondly to diversify the species used. The source of the raw material used for production has undergone substantial changes over the years. In the early days the main source of supply was the Philippines, and principally Mindanao. But when supplies from that area began to dry up, and with the imposition of a logging ban, supplies were obtained from Malaysia and Indonesia. Subsequently, the log export ban in Indonesia meant that supplies were almost entirely derived from Sabah and Sarawak.

In 1991, partly as a result of reports such as that by WWF (Nectoux, 1989) and the International Timber Trade Organization (ITTO) call to limit trade to timber from forests managed sustainably by the year 2000, the Japan Plywood Manufacturers' Association introduced a policy of reducing the use of tropical hardwoods in the manufacture of plywood. The policy was embodied in a set of non-binding targets in the 'Basic Direction of the Substitution of Plywood Feedstock' published in May 1991. The policy included targets to substitute about 30% of tropical timber with coniferous species by 1996 and to increase this further to a half in 2000. In a survey carried out by the same association in 1994 it was found that coniferous plywood accounted for 14.6% of domestic plywood production (Azuma, 1995) and in 1995 the level had reached nearly a quarter (Suzuki, 1996).

Not only have producers begun to respond to such factors, so too have consumers and other groups. Many cities and prefectures have published targets for the reduction of the use of tropical hardwood plywood in their public works programmes. For example, in 1991 Tokyo published targets to reduce tropical hardwood plywood to half of total usage by 1995 and to only 30% by 1997. Later plans were published for the use of domestic thinnings as a raw material.

However, Dauvergne (1997) argues that 'these policies have had few concrete results. ... [t]ropical log imports have decreased but this trend is the result of less abundant tropical log stocks, Southeast Asian log export bans, a shrinking Japanese plywood industry, and Japan's recession, not of greater concern for tropical forests'. He argues that, as most of the importing is done by the large trading houses, the effect is to reduce the price of tropical logs in Japan so mitigating against substitution by other species. The reasoning behind this is, first, the bargaining power of the trading houses and, secondly, the idea that they operate on very low margins, their aim being to maximize turnover rather than profits.

This assessment has some truth in it although it may be a little harsh. There has been an increase in the amount of coniferous timber used in plywood manufacture, with a substitution away from tropical hardwoods, although whether this is from altruism or self-interest is impossible to say. The latter seems more likely as it is certainly true that environmental concerns are of less interest to those who, in Britain, would be referred to as the 'chattering classes'. It is still far from uncommon to see tropical plywood being used as concrete forming for the most apparently trivial of items.

However, the trend in plywood manufacture to use coniferous logs began quite a while ago, with the attempts of the Akita plywood company in testing the production of plywood from larch from Russia in 1982 being an early example (Barr and Braden, 1988). The coniferous species mainly used have been radiata pine from New Zealand, Russian larch and to a lesser extent hemlock and Douglas fir from North America. However, as long as there were supplies of cheap, high quality tropical hardwoods it was not really worthwhile expanding production.

The views of producers of coniferous plywood were investigated by JAWIC (Japan Wood-Products Information and Research Center). A total of 37 factories, accounting for nearly a fifth of the output of coniferous-based plywood, were included in a questionnaire survey. Some of the results are presented in Table 2.5 and, if the perceptions in the table are general, it is clear that coniferous-based plywood has still a lot of ground to catch up if it is to be perceived as equal to that based on tropical hardwood.

A very small quantity of domestic timber is used. In the past this has been mostly hardwoods but recently sugi, karamatsu, ezomatsu and todomatsu have been used. However, the quantities have remained insignificant, still lying at less than 3% in 1996 - see Table 1.5. The problem here is reported to be distribution costs with the price required for felling the timber greater than the costs which producers are willing to bear, another indication that when Japanese timber competes in what might be called 'normal markets' it is not competitive on price (Azuma, 1995).

There also seems to be some resistance to the use of domestic coniferous timber, at least according to a survey of producers in 1993. Table 1.5 showed that karamatsu was used in plywood manufacture but this is seen to be somewhat inferior to Russian larch imports, especially in its use in concrete forming. Akamatsu was seen to have problems in gluing because of the large knot sizes and its strength was reduced if used in wet conditions such as concrete forming. Sugi was also not favoured because of the low strength of the plywood it formed and its poor characteristics in wet conditions. Also, the survey suggested that it

Table 2.5. Comparisons of hardwood and softwood plywood

	Coniferous plywood	Tropical hardwood plywood
Raw material	Annual rings often very visible. Knots and other faults are common and pre-planing is often necessary. Most logs are small diameter.	Annual rings almost indiscernible. Knots and other faults are few. Large logs available.
Technology	As the number of logs processed is large a high speed charger is needed. Large differences in the moisture content of heartwood and sapwood mean that drying is often necessary.	Conversion rate is high.
Product	Because of knots and other faults the surface is coarse and the appearance is poor. Warps and splits easily. Dimensional stability poor because of swelling and shrinkage under different moisture conditions.	Surface has a beautiful smooth finish. Dimensionally stable.
End-uses	Flooring, framing.	Furniture, building, shuttering, flooring.

source: Suzuki (1996)

should be avoided altogether in forming walls in housing, a problem for domestic producers if it is to increase its use in two-by-four housing (Doki, 1994).

However, despite these perceptions, coniferous plywood is gaining ground. A survey of 368 building companies showed that the numbers handling coniferous plywood had risen from 35% to 44% between 1993 and 1994 (*Mokuzai jouhou*, 2/95). The same survey revealed a growing acceptance of coniferous plywood for flooring and wall panels by the major house makers despite the comments about sugi above. Thus the likely increase in *bunjou* housing should create a better environment for the use of coniferous plywood although this might not extend to domestic conifers. But there was still a lot of resistance to its use amongst smaller builders, suggesting tropical hardwoods would still retain market share in the *mochi-ie* sector.

From the point of view of final consumers, there were apparently allergies reported early on because of the presence of knots. In this survey, rather disturbingly, it was reported that several end-users saw coniferous plywood as cheap and therefore bad. Clearly some careful marketing is required if it is to gain market acceptance and reduce the market for tropical hardwood plywood on the basis of price advantage.

Figure 2.10 shows the proportion of the total timber consumed in the plywood industry in Japan that was accounted for by non-tropical timber (i.e. all

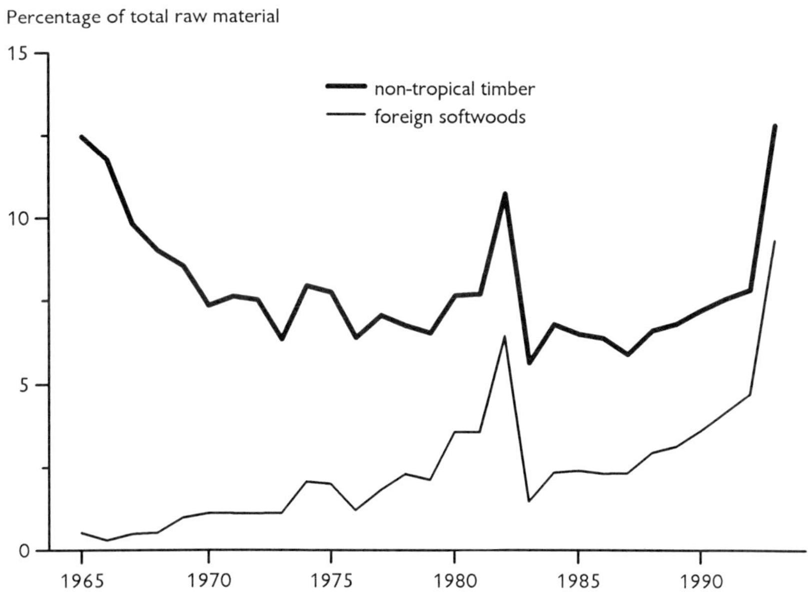

Fig. 2.10. Coniferous timber in plywood manufacture

of the timber not in the Japanese category *nanyouzai*). This has fallen and is now rising again but is not much different from the proportion registered in the mid-1960s. However, the dotted line shows the proportion of imported softwood in the raw material input. This has shown a distinct upward trend. The reason for the difference between the two lines is the level of the domestic timber, mainly hardwoods, used in plywood manufacture. This fell in the 1960s and 1970s.

The graph is possibly good and bad news from the environmental point of view. Whilst the amount of tropical timber finding its way into Japanese plywood is falling the use of Russian timber has its own ecological problems and the use of radiata pine and North American timber is also not without its ecological issues.

Other panel products

According to Japanese figures, the plywood market accounts for about three-quarters of the panel products market. Figures for 1994 show that the overall panel products market could be characterized by Table 2.6.

The figures for plywood are not identical with those in Fig. 2.9 as slightly different conversion factors seem to have been used, but the dominance of plywood is apparent. However, the Japanese do have a long history of particle board production with the patterns of output looking very similar to those of plywood.

Table 2.6. Other panel products

	Plywood	Hardboard	MDF	Insulation board	Particle board	Total
Domestic production	4,900	120	330	640	1,220	7,200
Imports	4,100	5	420	20	350	4,895
Total	9,000	125	750	660	1,570	12,095

All figures in thousands of cubic metres.
source: Himeno (1995)

The use of these panel products has been affected by the relative cheapness of plywood and the visual characteristics of the various panels, none of which are really suitable for traditional items in housing or for use as concrete forming. The demand for these products is, however, now increasing with the trends in housing outlined above. The present end-uses of the various panel products are outlined in Table 2.7.

Table 2.7. End-uses of other panel products

End-use	Hardboard	MDF	Insulation board	Particle board
Construction	12.8	37.0		31.3
Furniture	17.3	18.6		55.5
Car manufacture	34.8			
Packaging	25.7			
Electrical equipment	3.6	3.3		10.3
Ceilings and walls			8.1	
Core for laminated panels			20.9	
Flooring			62.6	

All figures are percentages, other uses account for the remaining usage.
source: Himeno (1995)

Concluding remarks

The market for panel product depends largely on levels of economic activity in the general sense. Their use in housing will depend mainly on the increase in *bunjou* type housing and the reduction of traditional forms in the *mochi-ie* sector. The increasing cost advantages of panel products mean that plywood will continue to be substituted for solid wood. Not only will plywood gain market share but so too will other forms of panel products as cheaper, mass-produced housing increases.

The survey referred to earlier (Suzuki, 1996) reported that of the output of domestically produced coniferous plywood, 16.5% went to producers of prefab housing, 37% to producers of two-by-four housing but 23.7% to post-and-beam housing. Thus, to the extent that *mochi-ie* is predominantly post-and-beam, an increase in demand here will also cause an increase in plywood demand. However, the stable or falling housing starts predicted in Table 2.2 could still imply increasing demand for panels if replacement housing demand substituted two-by-four or western-style housing for traditional styles. What does seem certain, however, is that demand in general will rise and coniferous inputs will increase.

Pulp, Paper and Paperboard

The Japanese have a long history of paper making but the modern paper industry was set up after the Meiji restoration as one of the pillars of western-style industrialization. Since then the industry has expanded and Japan now has a very high level of per capita paper consumption, presently standing at fifth in the world with a figure of 248.7 kilograms according to the *Pulp and Paper International Annual Review*. This ranks only behind the USA, the world leader with a consumption of 334.6 kilograms per capita, Finland, Sweden and, rather strangely, Belgium. The British figure is only 208.2.

The consumption of paper and paperboard is shown in Fig. 2.11 and a significant feature of the trend is that, while the break in the time series for the first oil shock is clear, after that demand resumed its upward trend, unlike that for other timber products.

This demand has been largely supplied by imported chips and pulp and the self-sufficiency ratio in RWE in the pulplogs and chip category has recently

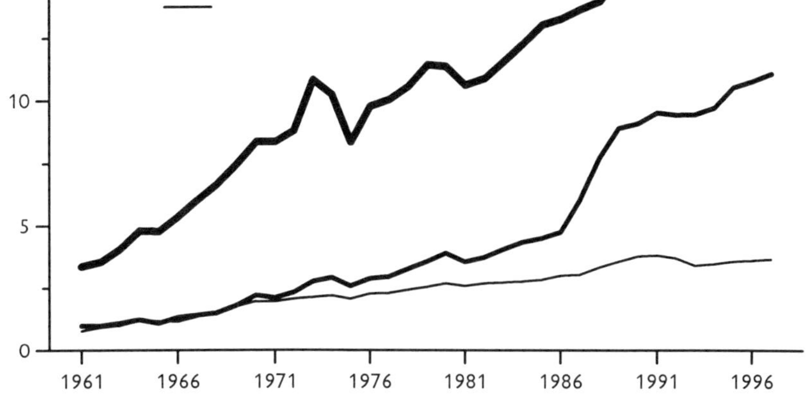

Fig. 2.11. Consumption of paper and paperboard

fallen considerably from 21% in 1992 to only 12.9% in 1996. The deciduous species found mainly in natural forests have provided an important source of raw material for this industry. The main domestic source of pulp and chips has been in the category of broad-leaved species which the Japanese classify as 'other', and now over 80% of this category is used for pulp production. However, the use of nara and buna for pulping has also, in the proportional sense, been increasing. Both of these trends have largely been caused by the drop in the use of hardwoods for lumber. In the mid-sixties around 70% of nara and buna was used for lumber, as was around 30% of other deciduous species.

The use of deciduous and coniferous species for pulp is shown in Fig. 2.12. The coniferous species are mainly those other than sugi and hinoki, although a little sugi is used in the pulp industry.

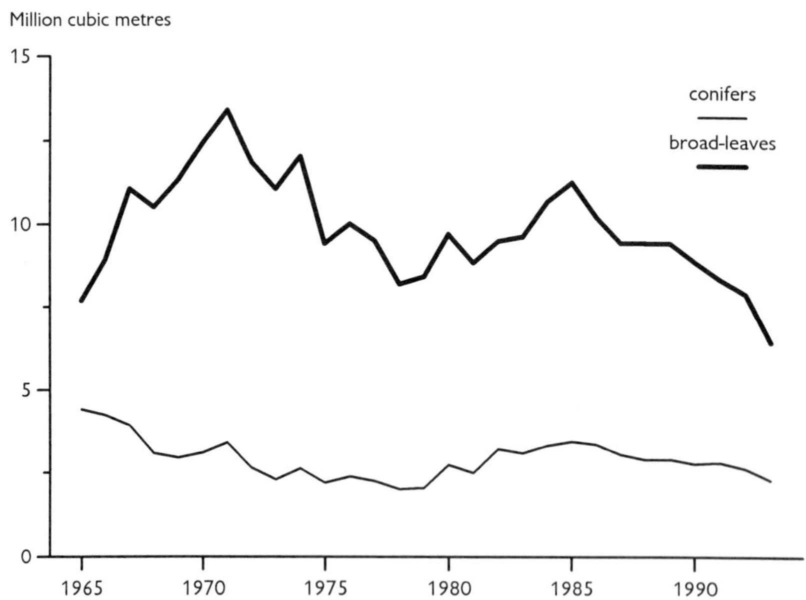

Fig. 2.12. Domestic supply of pulpwood and chips

In recent years the amount of pulp coming from domestic forests has been falling. Three reasons can be given for this, one is that the large producers of paper in Japan operate in a traditional Japanese *keiretsu* manner which, it is claimed, transfers profits to the main companies by depressing the price of raw materials and intermediate products. This reduces the price of domestic chips and pulpwood. A second reason is the cheapening price of chips from abroad. And finally, of course, a lot of the supply of chips and pulpwood emerged when natural forests were replaced by man-made ones and, with the stagnation in the lumber market, there is less of an incentive to convert natural forests into man-made ones (Koide, 1992).

One feature of the industry is the reduction in the use of wood for pulp and the increase in the amount going for chipping. In the mid-sixties around a third of other hardwoods entered the end-use classified as pulpwood. Now the figure is only around 1%. This represents a rationalization of the paper industry from, relatively speaking, a small scale operation to a more efficient, integrated one.

An important factor is the use of recycled paper in the production process. Japan has high rates of waste paper collection and recycling, although these have stabilized in recent years. Around 54% of the furnish for paper production consists of recycled paper. Recycling is promoted as being able to deliver environmental benefits as well as a means to reduce the levels of waste that have to be disposed of by other means. Also, Japanese paper makers and trading houses which are involved in the import of chips for the paper industry are coming under increasing pressure and so high levels of recycling are seen as ways of promoting a better 'image' for the companies concerned.

In 1995 the Japan Paper Association set an industry target of 56% recycling by the year 2000. This is still below the level which the Association reports is technically feasible if current technical standards are to be maintained. This level is nearly 70%.

The main factor affecting recycling is probably the costs involved in collecting the paper compared with the costs of using virgin fibre. With the fall in the value of the yen the relative prices of these two fibre sources will probably move in favour of recycling and so increases in recycling rates might be expected.

Conclusions

The brief description of the major demand patterns and changes in the Japanese timber market has, of necessity, concentrated on the housing market. As a very broad generalization, the long history of Japanese forestry outlined in Chapter one and the period of isolation of the country under the Shogunate created a forest industry which had a number of peculiarities almost all of which were related to the nature of traditional housing. This is especially true of the market for *migaki-maruta* or logs that make up the polished pillars around the *tokonoma*, or alcove, in traditional Japanese rooms. Here, pieces of bamboo or plastic are wrapped around the stem of the tree and left there to create a wavy surface on the debarked column which has a pleasing ornamental effect.

In the years after the Meiji Restoration there was a widespread adoption of western styles, including in housing, but it was in the early 1970s when the ordinary Japanese began to travel that the acceptance of, and indeed preference for, western-style housing really took off. The writer remembers being somewhat surprised when first talking to officials at the Forestry and Forest Products Research Institute to discover that most of them actually preferred western-style housing to traditional Japanese forms.

Here, one is reminded of the apocryphal tale of British forests being full of useless oak trees specially formed for timber for shipbuilding when technology had moved on to iron ships. In Japan, forests and forestry designed for a bygone age of traditional housing are having to meet the challenge of imports from countries where costs are lower and the timber can meet the increasing technical rather than aesthetic demands of the changing market.

Thus, the new methods of construction, the demand for panel products and pulp and paper are all areas where Japanese markets are becoming more like those in other countries. There seem to be no areas where the potentially increased supply of Japanese timber can find a market which would support a higher price than that of imports. In some parts of the country there are attempts to market ‘local timber’ in housing but the trends in the housing market make it clear that economic factors, either of the price of the timber itself or the costs associated with producing traditional styles of housing, are stimulating the increase of modern housing methods. If this is not a gross caricature of the Japanese market it means that the level of imports will increase further with the self-sufficiency ratio in the lumber market, the last sector in Japan to maintain a significant domestic input, falling to the levels recorded in the plywood and pulp sectors.

However, foreign sources of timber are themselves facing problems so it is no easy matter to merely extrapolate the trends seen in this chapter. If foreign supplies start to dry up the domestic supply of timber could well increase if prices increase to reflect scarcity. To examine these possibilities it is best to remain in Japan, for one more chapter at least, and look at the domestic supply of timber and its likely development in the future.

Chapter three:

The Domestic Supply of Timber

Introduction

The increasing reliance of Japan on imported timber was described in Chapter one. So too was the apparent paradox that Japan is home to an increasing stock of maturing timber. The significance of this, of course, lies in the fact that many of the sources of timber for the Japanese market are themselves coming under increasing pressure. In America and Southeast Asia the level of timber output is not likely to expand significantly in the near future. In Russia many different problems are influencing the level of forest products production. Only in New Zealand and Chile do increasing timber output levels seem likely.

The figures from the Forestry White Paper for 1997 give a total growing stock for Japanese forests of 3483 million cubic metres. This is around 30 times the annual demand for timber in Japan. If the potential output which this figure represents is compared with the actual production figures, which give Japan a self-sufficiency ratio of only 20%, it is plain that the biological feasibility to increase supply is enormous.

An important question, therefore, is the degree to which domestic supply can be substituted for imported timber and timber products. There are really two questions here. One is the technical question of whether Japanese timber can be freely substituted for imported timber. The second question is whether Japanese timber will actually come onto the market. In a sense, these are different aspects of the same question. If the price of imported timber were to rise sufficiently there is little doubt that Japanese timber *could* be used for most end-uses. Some of these trends are already apparent with the increasing production of coniferous based plywood. However, these questions will not be examined in this chapter. Here, the maturing timber stocks referred to in Chapter one will be the subject of discussion. These have the potential to play a very significant role in the domestic timber market, and therefore influence the international timber trade.

From the point of view of forecasting, they also pose a problem in predicting the future course of the Japanese timber market.

With this in mind, therefore, it is necessary to look in some detail at the Japanese forestry sector to determine the reasons for felling (or not felling) and, from the analysis, to derive a likely set of forecasts. This chapter will look at the first of these areas, the nature of Japan's forestry production. The Japanese forecasting methods will also be briefly outlined but the forecasts themselves will be the subject of Chapters seven and eight.

The Output of Timber from Japan's Forests

In Chapter one it was shown that the output from Japan's forests had fallen as the country was opened up to trade in timber products and foreign sources of timber had proven cheaper than domestic ones. However, the fall in output has been different for different species and this might be a useful starting point in the analysis of Japan's potential domestic output.

The decline in output of total domestic timber was shown by the declining self-sufficiency ratio in Fig. 1.3 in Chapter one. The output of coniferous timber in 1996 was only 51.5% of its 1965 level and the drop in output of the various coniferous species is shown in Fig. 3.1. The line labelled 'other conifers' contains a number of species, but of them karamatsu, ezomatsu and todomatsu

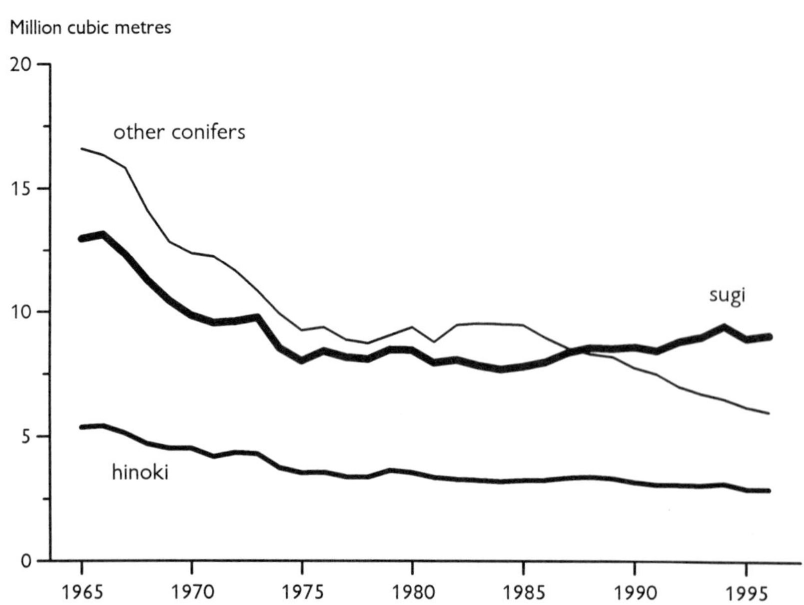

Fig. 3.1. Coniferous log output

have shown output levels which have declined roughly in line with sugi and hinoki while akamatsu, kuromatu, momi and tsuga have shown the most marked drops in output.

An interesting feature of the graph is the recent upturn in sugi output. It is this which sometimes causes commentators to comment that an era of self-sufficiency is at hand. However, the output of hinoki seems to show a consistently downward trend. Given the price which hinoki can command this is surprising as it would seem likely from an economic point of view that, if an increase in domestic log output were to occur, it would occur to hinoki first. The downward hinoki trend and upward pattern for sugi probably suggest that the market is becoming less 'Japanese' in nature. Sugi is competing on price and holding its own in the market but the market for high quality hinoki does not have sufficient depth to sustain both an increase in output and a high price premium.

The changes in the end-uses of the coniferous species show a consolidation of their use for lumber and the beginnings of their use in plywood production. These trends are shown in Table 3.1. The first column in the table shows the drop in output from 1965 to 1996, for example the output of sugi fell by 69.9% over the period, from 12,984 thousand cubic metres to the 1996 figure of 9078 thousand. However, its use in lumber fell to only 72.8% of the 1965 figure, from 12,006 thousand cubic metres to 8738 thousand. Therefore, relatively speaking, the use of sugi for sawnwood increased over the period and its use for other purposes declined.

Table 3.1. The changes in the end-uses of coniferous species

	Total	Sawnwood	Plywood	Pulp	Other
All conifers	51.5	54.3	*	51.8	14.2
Sugi	69.9	72.8	*	127.6	14.3
Hinoki	56.8	56.1	*	*	47.9
Akamatu & kuromatsu	27.1	22.6	*	42.7	12.3
Karamatsu, ezomatsu & todomatsu	66.2	77.3	-	61.2	*
Momi & tsuga	11.6	10.3	-	21.3	16.4

* The original usage was zero but some use is now recorded.
- No usage registered for first or final year.
All figures are percentages.
source: various editions of *Mokuzai jyukyou houkokushou*

In general the table shows that the use of sugi, karamatsu, ezomatsu and todomatsu for lumber did not fall by quite the same amount as the drop in their overall output but their presence in the 'other' uses category dropped by more. For akamatsu and kuromatsu there was a substitution in the other direction away from lumber and into pulping, as there was for momi and tsuga. These figures imply that the present end-uses shown in Table 1.5 in Chapter one have been changing slightly, with the main difference occurring in momi and tsuga.

The output from natural forests exhibits a somewhat different pattern and this is shown in Fig. 3.2. Unlike the continuous decline in output of coniferous logs shown in Fig. 3.1, hardwood output peaked in the 1970s and only began to decline afterwards. This was mainly as a result of the increase in demand for pulp and chips for which the 'other hardwoods' were used. Nara and buna have only entered the pulp market relatively recently and so their output levels did not exactly mirror this trend.

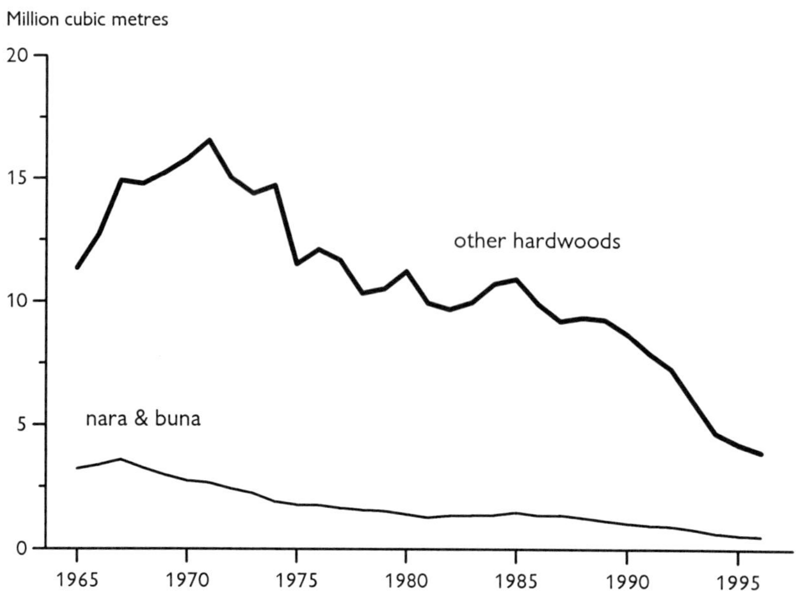

Fig. 3.2. Hardwood log output

Total hardwood log production in 1996 was 30.7% of its 1965 level, a greater drop than that of the conifers. However, again there has been a shift away from lumber and, for nara and buna at least, a shift into plywood use which is shown in Table 3.2. The 'other hardwoods' are now predominantly used in pulping. Thus, overall, the value of the end-uses of hardwoods has fallen with both nara and buna being increasingly used for chips.

Two reasons for the decline in output shown in Figs 3.1 and 3.2 might be suggested. One is the drop in profitability which goes with the changes in the rural economy outlined earlier. Another could be the effect of changes in the objectives of forest managers away from profit toward other environmentally oriented outputs. It is possible that, if national and private forests are operating under different criteria, with private forest owners being more financially oriented, the output trends of these two groups will be different. It might be useful, therefore, to look at the output of Japan's forests according to source.

Table 3.2. The changes in the end-uses of hardwood species

	Total	Sawnwood	Plywood	Pulp	Other
Nara	44.1	29.0	163.2	*	0
Buna	17.3	14.9	49.6	17.5	2.6
Other hardwoods	52.7	21.2	33.2	80.1	10.6

* The original usage was zero but some use is now recorded.
All figures are percentages.
source: various editions of *Mokuzai jyukyou houkokushou*

Trends in Output from National and Private Forests

Figure 3.3 shows the total log production from the three main categories of forest, national, public and private. The immediately obvious feature of the graph is the decline in output from all three sources. However, in the early period at least the decline in output from private forests was more marked than that from the other two categories. In recent years, however, this pattern has been reversed with private forest output holding up somewhat better than that from the national forests.

In proportional terms, the output from public forests has stayed fairly constant at between 5% and 6% of the total output. National forest output accounted for around 22% of output at the beginning of the data set shown in

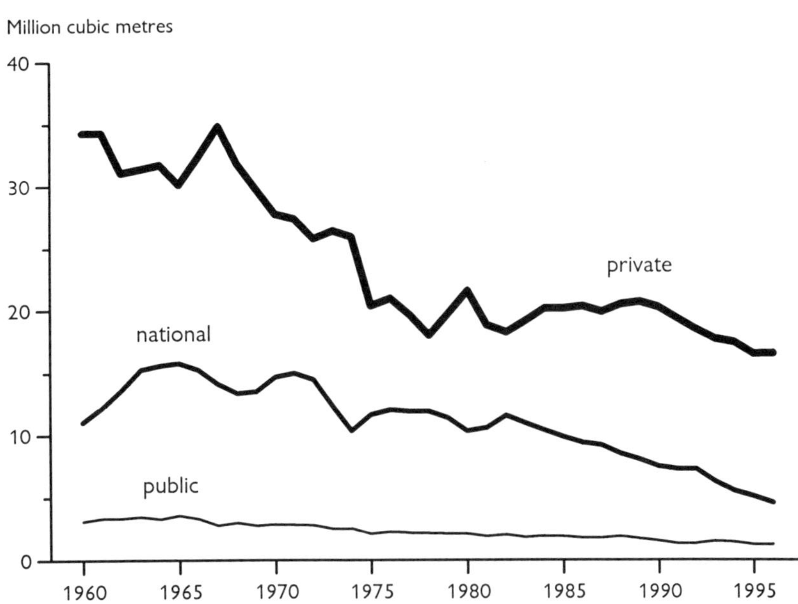

Fig. 3.3. Log production from Japan's forests

Fig. 3.3 and now accounts for an almost identical proportion. However, the figure rose during the 1960s and 1970s to around 35% and then began to fall back to its present level. The figures for output from private forests followed, of course, the opposite pattern, falling to around 60% before rising again.

In a sense, the fact that the proportion of timber originating from private and national forests has remained roughly the same is somewhat surprising as the nature of the forests shows some interesting differences. It has already been seen in Chapter one that the national forests contain a greater proportion of natural forests and, therefore, a greater proportion of hardwood species. As these have shown an output pattern somewhat different from that of the coniferous species this ought to be reflected in the overall figures.

However, the bulge in output from national forests seen in Fig. 3.3 does not seem to be accounted for by extra output of broad-leaved species for pulp. Output of conifers more-or-less mirrored that of hardwoods during the period. National and private forests have somewhat different locations. As a general rule, the elevation of national forests is higher and their slope is greater than private forests. This is shown in figures from the Forestry White Paper for 1996 which are reproduced here in Table 3.3.

Table 3.3. The location of national and private forests

	National forests (000 ha)	Private forests (000 ha)
Elevations less than 400 m	2,196	10,381
Elevations from 401 m to 1000 m	3,922	5,883
Elevations above 1000 m	1,726	1,038
Slopes less than 15°	1,883	5,191
Slopes between 16° & 30°	3,137	7,613
Slopes greater than 30°	2,824	4,498

source: adapted from Forestry White Paper (1996)

The slope figures need to be interpreted with reference to the figure quoted in Chapter one that only 30% of Japan's land has a slope of less than 8°. Most of this land is for urban and agricultural use and so the category of less than 15° in Table 3.3 probably means between 8° and 15°. There is a slight preponderance of private forests on the flatter sites. In fact, 73% of the forests on the lower slopes are private compared with only 61% on the slopes over 30°. This clearly has implications for the costs attached to forestry and the ability to increase labour productivity by the introduction of modern harvesting technology. Most Japanese experts quote a figure of around 15° as the limiting slope for most mechanized systems (with the obvious exception of winch systems) to be practicable. If this is true it means harvesters, forwarders and the like will only be of use in less than a third of Japan's forests. The elevation figures show a greater disparity. About 82% of the forests lying at altitudes below 400 metres are in private hands whereas 62% of the forests at higher altitudes are national.

This disparity in location was probably greater in the years prior to the

Second World War. One of the features of post-war forestry in Japan has been the movement of man-made forests, and therefore the important coniferous species, 'up the hill'. This is especially true in southern Honshu and Kyushu. So much of the private forests in the less favourable locations will be of fairly recent origin.

National forests

The policy relating to national forests has essentially shown a gradual trend away from timber production towards the use of forests for environmental and other non-timber purposes. However, despite the early forest laws and policies which were aimed at establishing something like a sustained yield output from Japan's national (and private) forests, the output of national forests in the 1950s and 1960s exceeded the annual increment, often by a factor of two (Ueki, 1992; Kumazaki, 1997).

Originally, against the background of the high demand seen in Chapter one, the development of forests of the interior was accelerated by the expansion of roading to create an infrastructure for the logging and processing industry. Accompanying this was a policy to convert broad-leaved forests to more desirable and productive coniferous ones. After the oil shock of 1973 and the increasing competition from cheaper imports the policy review of that year stated:

> 'Forest resources have both a timber product side and a land preservation side as well as possessing the public functions of preservation of natural resources, development of water resources, etc.' (Ueki, 1992)

Thus, in the same way that multiple-use forestry was making inroads into forestry in the rest of the industrialized world, so too in Japan the value of other functions of national forests was beginning to be felt. However, despite these words, the targets for the conversion of natural forests to man-made ones continued to be increased, partly as a response to the second oil shock in 1978 when imports seemed likely to drop.

Coupled with forest policy were calls to use the forest industry as a means to revive the rural economy. But, as Fig. 3.3 shows, the output of Japan's national forests has been falling and so their contribution to local economies and employment has mirrored this trend. At the same time, the growing stock has been reaching maturity so that now output has fallen to only half the timber increment. Overall, the national forests are running a substantial deficit and this means that there are now moves to reorganize the Forestry Agency. Whilst there do not seem to be the calls for total privatization as there have been in other countries, indeed it is doubtful that anyone would want to buy Japan's national forests, the idea of moving part of the Forestry Agency to an environment ministry has been floated.

Thus, despite the increasing growing stock, it is unlikely that national forests will lead a move to supporting a domestically-based forest industry. Indeed, any move towards that end by means of subsidies or the like would bring accusations of 'unfair' competition. The thinking behind the suggested move of the Forestry Agency to an environment ministry suggests that, at last, the non-timber uses of national forests are dominant

This seems to be in line with the avowed wishes of the Japanese themselves. Although it is difficult to make comparisons between different questionnaires through time, the expectations for forests are surveyed fairly regularly by the Japanese. One feature of all of the questionnaires is that, when asked what they expect from Japan's forests, only a fairly low proportion answer 'timber production'. This proportion has fluctuated but does not seem to increase significantly. The figure was 19.3% in 1984 (*Shinrin keikaku kenkyuukai*, 1987), 27.5% in 1998 (Forestry White Paper, 1996), 26% in 1993 (Forestry White Paper, 1994) and 22% in 1996 (*Rinyachou*, 1997). The figures also suggest that the feeling that forests are for timber production is strongest in the older age groups, reaching 38.2% in the over-seventies.

Partly as a result of this, and in line with the discussion above, the new forest plans look to a slight decline in the area of forest where timber production is one of the main objectives. The proposed area allocations for Japan's forests are shown in Table 3.4 and, although the figures apply to both national and private forests, they are a reflection of government thinking.

Table 3.4. The target areas for various forest functions (000 ha)

	1980 plan	1987 plan	1997 plan
Timber production	17,570	15,800	14,900
Watershed protection	11,810	14,350	14,670
Mountain slope protection	3,660	4,360	5,890
Amenity	3,320	3,550	4,320

Because of multiple uses being attributed to forest areas the sums of the columns do not correspond to overall forest areas.

sources: MAFF (1980, 1996); *Rinyachou* (1997)

However, despite the discussion of environmental factors, the output of national forests seems largely to have been determined by the economics of timber production. Not only has output been falling but it has consistently been below the level planned. This discrepancy seems to be increasing. In 1970 the volume felled was 95% of the level of felling planned by the Forestry Agency, but in 1990 this had fallen to just over 80% (Iida, 1992). Indeed, some might suggest that the slow movement towards environmentally based management plans has been forced on the Forestry Agency by the low price of timber.

It might be thought that the reduction in the area allocated for timber production would be directly factored into the forecasts of timber output or that the features of government policy would relate directly to the prediction of timber production. However, as will be seen in Chapter seven, this is not so. Indeed, the output of national and private forests are forecast in the same way. The conclusion embodied in the Japanese forecasts seems to be that much of what determines the output of private forests also impinges directly on national ones, that is timber prices, costs, etc.

Private forests

The reduction in output of private forests has been the most marked in absolute terms, although the proportional fall in output is similar to that for national forests. There are two general reasons that could account for this fall. The first would be a change in the composition of private forests, for example a reduction in their overall area or a movement in their ownership from owners who actively manage to those who did not. Secondly there might be a general decline in the level of production from all of the subgroups which make up private forests. As will be seen, the latter appears to be the dominant force at work in Japanese forests.

The category of *shiyuurin* in Table 1.1 in Chapter one is usually translated as private forest but it contains a number of sub-categories. The most detailed figures for these categories groups obtained in the decennial Forestry Census and the latest figures, from 1990, are given in Table 3.5. The main sub-category is 'households' which is also itself broken down in more detailed presentations into agricultural/forest households (*noukarinka*) and non-agricultural/forest households (*hi-noukarinka*). These groups hold the largest area of forest, followed by companies and communal ownerships.

Households

The household group is clearly a large one but one which has a very small average holding. Over the period from the first census in 1960, the average size of household forests has increased from 2.4 hectares to the present figure of only 2.7 hectares. This figure hides a typically skewed distribution of ownership

Table 3.5. Private forest ownership

	Number	Area (ha)	Average area (ha)
Shiyuurin	2,859,492	10,682,879	3.7
Households	2,508,605	6,752,121	2.7
Companies	43,937	1,520,293	34.6
Temples & shrines	33,628	152,598	4.5
Joint ownerships	202,786	710,140	3.5
Associations and co-ops	11,327	405,435	35.8
Communal ownerships	59,209	1,142,292	19.3

source: Forestry White Paper (1996)

size which is shown in Fig. 3.4.

The dominant size of holding is in the 0.1 to 1 hectare range with 57.9% of households falling into this class. As would be expected, this small scale of holding mitigates against intensive forest management. The fixed costs of harvesting; planning, setting up extraction systems, marketing, etc., will all have to be spread over a felling area of very small size. Not only that, but felling will be a once-in-a-lifetime activity and the income received very small. All of this means that the likelihood of felling, all other things being constant, will be greater for larger holdings of forest.

This is borne out by the data, but more worryingly for future output of domestic timber, the felling intensity for any scale of holding is falling. The Forestry Census asks if, during the previous year, particular activities have been undertaken. Naturally, one would expect the probability to be higher for the larger holdings as, given the greater likelihood that the larger area encompasses forests of different age classes, felling, thinning and planting are more likely to be undertaken in any given year. The results of the previous four censuses are given in Table 3.6. Only those households with forest holdings in excess of one hectare are included in this part of the census. Thus, in 1960 some 26.9% of all households sold timber and this figure has fallen to only 5.2% in the latest census.

As would be expected, the sale of timber in the year previous to the census is greater in those households with a larger forest holding. However, in terms of

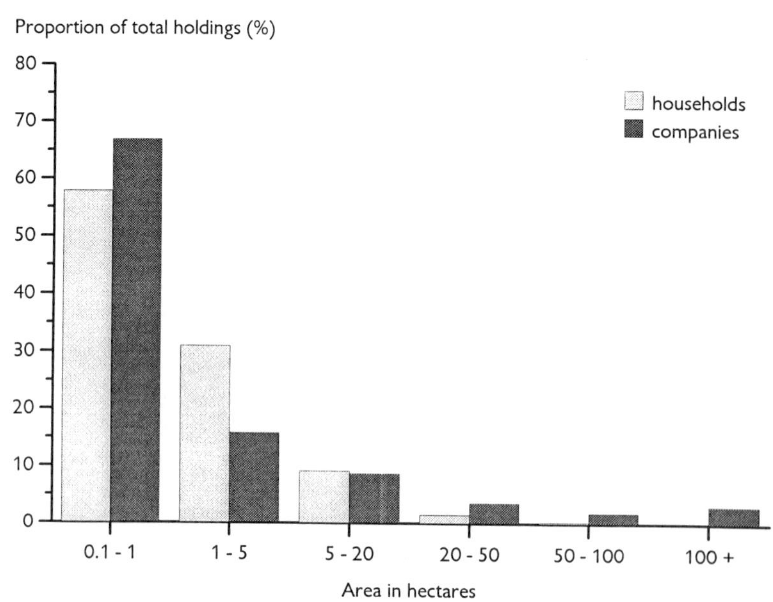

Fig. 3.4. The scale of forest private holdings

Table 3.6. The management intensity of households

		Scale of holding (%)				
		1-5 ha	5-20 ha	20-100 ha	100+ ha	Average (%)
Sale of timber	1960	20.4	46.7	62.0	65.7	26.9
	1970	12.5	31.0	52.4	62.5	18.1
	1980	4.0	12.0	28.1	45.3	6.9
	1990	2.8	9.0	22.6	36.0	5.2
Tending	1960	52.8	68.3	78.2	77.5	56.6
	1970	56.1	71.5	81.3	80.9	60.4
	1980	41.2	54.5	66.3	66.5	45.2
	1990	24.0	34.4	45.6	46.3	27.3

source: Akaha (1992)

management intensity, it might be possible to argue that the figures for the smaller households show somewhat greater felling activity. For example, if a household had 5 hectares which consisted of five 1-hectare forests with a rotation age of 50, felling might be expected to occur, say, once every 10 years. Thus, the 20.4% in 1960 would represent a very high level of felling activity. On the other hand, if a holding of 100 hectares had a similar age class structure felling would be required every year, giving a figure of 100% in Table 3.6.

The part of the table marked 'tending' implies some form of forest operation such as pruning, snow clearing, cutting of climbers, etc., in fact, any operation in the forest other than thinning, planting or felling. These figures are a lot higher than the sale of timber, as would be expected, and do not really suggest a lot of the forests are being totally neglected with the proportion of forests being tended holding up quite well at least until the final census. Also, the age of forests in Japan is increasing and so a decline in these activities, which are concentrated in the early part of the rotation, might be expected.

One of the reasons underlying these drops in activities may be the increase in the number of absentee forest owners, i.e. those who do not live near their forests. This trend will increase the cost of management and, also, a forest owner living away from the forest is more likely to be engaged in some other form of work and so not interested in forestry.

Although the figures do not appear to distinguish between the various categories of private forests, the number of absentee forest owners, defined as those who do not live in the same town or village as their forests, has increased from 14.9% in the 1970 census, to 21.8% in the latest one in 1990 (Akaha, 1992; Tanaka, 1997b). Also, the number of these who live outside of the prefecture, rather than just the town, has risen. The areas where the largest numbers of absentee forest owners are recorded are in Nara, Wakayama and Hokkaido. However, these figures might be a manifestation, paradoxically, of the more industrial nature of forestry in these areas, with distant forests being used for timber production. However, the high figures which are also noted in the Tokyo area, Kanagawa, Gifu, Gumma, and Shizuoka, all of which are within reasonable driving distance of the Tokyo metropolis, betray the fact that some of these forests were purchased speculatively during the 'bubble economy'.

Much of the government's policy towards private forests has hinged on the establishment and encouragement of forestry co-operatives which should be in a position to take over the management of such forests. Now, co-ops operate auction markets and some have their own sawmills and market the lumber cut from their members' forests. They are very strong in the areas of planting and tending and, through this activity, act as a channel for subsidies from local governments to forest owners (*Zenkoku shinrin kumiai rengoukai*, 1991; Elwood, 1992). Forestry co-ops are now responsible for nearly three-quarters of the planting and about 15% of log production from private forests (Forestry White Paper, 1997). However, Chapter one showed the problem of the ageing work force in such co-ops and the production of the co-operatives is showing a similar decline to that shown in forestry as a whole.

Companies

Figure 3.4 also shows the distribution of sizes for company forests. Somewhat surprisingly, apart from a few large-scale owners, the company sector does not seem to own forests that are noticeably larger than those of households. However, this is a relatively recent feature. The figures from the earlier censuses show that between 1960 and 1970 the average size of company holding dropped from 224.9 hectares to 90.3 hectares. Since then it has continued to drop to reach its present level of only 34.6 hectares (Akaha, 1992).

Part of the reason for this is the purchase of forests for purposes other than forestry. The figures show that by far the largest increases in the number of companies holding 1 hectare or more has come in the construction, real estate and service sectors. The former has increased from 408 companies in 1970 to 2258 in the latest figures, real estate and financial companies have increased in number from 884 to 2646 and the service sector has seen an increase over the same time period from 533 to 2833. At the same time companies classified as being in 'forestry' rose from 356 to only 667.

Given the relative illiquidity of the market for forest land it is unlikely that the demand for forest land by financial companies represents some sort of portfolio diversification as has been seen in, for example, the activities of pension funds and insurance companies in Britain. The reason for this non-forestry based activity is mainly the desire to redevelop forest land for other uses such as golf courses and country clubs. The size distribution of the holdings of these non-forestry enterprises is at the smaller end of the spectrum and, therefore, serves to bring the size of the average holding down and contribute to the preponderance of ownership on the 'left' of Fig. 3.4. On the other hand, the forestry-related companies have holdings concentrated in the upper end of the distribution of sizes, showing an average holding of around 150 hectares.

The distribution of company forests around Japan is also significant. They tend to be concentrated in Hokkaido, where they are mainly large in size and concerned with timber production, and around the Kanto plain area, in the vicinity of Tokyo and the other major urban centres. Here they are relatively small and the main aim is one of redevelopment.

The level of management intensity is not particularly great for many of these companies and so it would be a mistake to think that forest moving from the household to the company sector would come under rational forest management and so contribute to supply. Many of the service and construction companies do exhibit high rates of felling but these are also associated with low rates of selling forest produce. For example, 36.4% of the companies in the service sector recorded felling activity in the year of the last census but 97.8% also recorded that they had sold no forest products. This suggests that the timber felled to clear sites for redevelopment is not sent to the market. Given that the only costs that would be associated with the marketing of such timber are felling and transport, this suggests a very low level of profitability for forestry.

Concluding comments

From the discussion above it seems that the slight changes in the average size of private holdings and the movement from households to company ownership are not significant in terms of trying to explain the overall fall in output. This is a somewhat disappointing conclusion. In North America and Scandinavia, the characteristics of forest owners, and the changes in them, have been related to felling intensity and log output - see for example Dennis (1989) or Loikkanen *et al.* (1986). The author has tried to reproduce these studies using Japanese data, but the results have been disappointing. The age of forest owners was found to be positively related to felling intensity, a rather strange result as in other studies it is usually the younger forest owners who fell timber. The felling intensity also appears to be negatively related to the distance from home to forest, suggesting that the increase in absentee forest owners could be part of the explanation for falling output. But these relationships seem to be weak and so explanations for the falling rates of management intensity have to be sought elsewhere (Blandon, 1991a).

In any case, the same drop in output has been seen in national forests with no change in ownership. So, if the change in private forest ownership patterns had provided a plausible explanation of the fall in felling activity, a separate explanation of the drop in output in national forests would have been needed. If national forests have been managed largely for profit, which the previous discussion suggested, it would seem that the factors which are causing the fall in log output are acting more-or-less equally on national forests and private ones. In this case, it might be useful to look at the economics of forestry in general, with special reference to man-made forests.

Man-made Forests

To begin the discussion, it is useful to look at the areas of timber felled in the various age classes in man-made forests. This gives an insight into one of the problems associated with forecasting the output from man-made forests. Figure 3.5 shows the areas of man-made forests of all species which were felled during the period 1990 to 1995 and is taken from data for the age class structure found in Forestry White Papers. The 'felled' area is simply calculated by subtracting the standing area in one age class from the area in the preceding age class 5 years earlier. This may not be a very accurate way of calculating the area felled but the overall picture given is realistic.

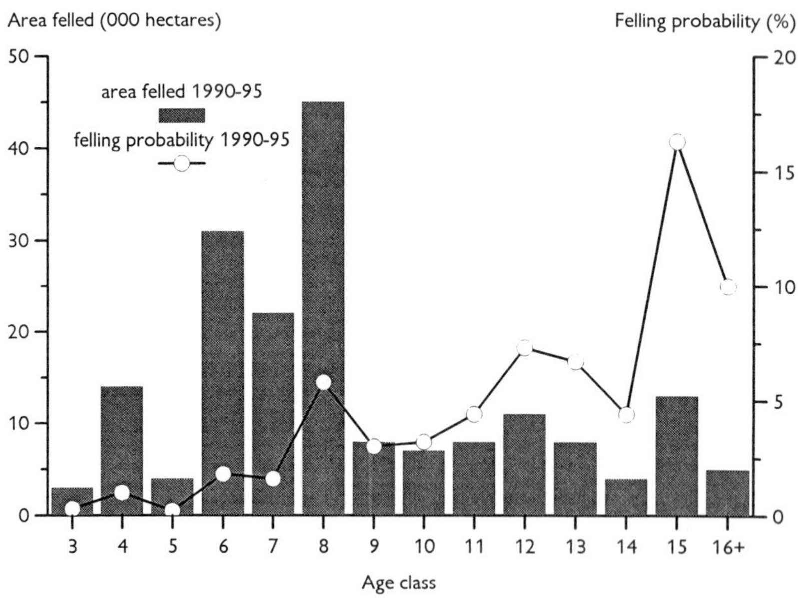

Fig. 3.5. The felling behaviour of owners of man-made forests

The immediate impression from the diagram is that the range of felling ages is very wide. This is in contrast to the usual textbook approach to determining the optimum rotation age of a forest which is characterized by the Faustmann formula and its extensions. One of the features of the Faustmann formula is the fact that, in most situations, the results it gives for the optimum felling age are fairly invariant to changes in the variables which are used. In other words, it is difficult to explain the wide range of felling ages in the diagram in terms of local differences in prices, costs or growth rates.

In Fig. 3.5 the 'felling probability' is presented. This is merely the area felled divided by the original standing area in the age class. Thus, the high area felled in age class 8 is less impressive when it is seen to be only 5.8% of the standing area in that age class. The felling probability is a pivotal concept in the Japanese attempts to forecast the output from private forests and is also central to the models developed here. The probability shows a general rise with age class, as would be expected. However, the figure for age class 16+, which corresponds to stand ages greater than 80 years, is only 10% which implies that a large proportion of the stands in Japan are being felled at very late rotations.

The felling probabilities themselves have been consistently declining through time reflecting the drop in output. This is shown in Fig. 3.6 which gives the felling probabilities for the two 5-year periods 1985 to 1990 and 1990 to 1995. The probabilities for the later period are the same as those in Fig 3.5 except that the series is not as detailed. The area figures for all man-made forests reported in 1985 were not as complete as they were in later years. However, the situation is clear and it is one which is repeated across all prefectures and all species. The propensity to fell is dropping.

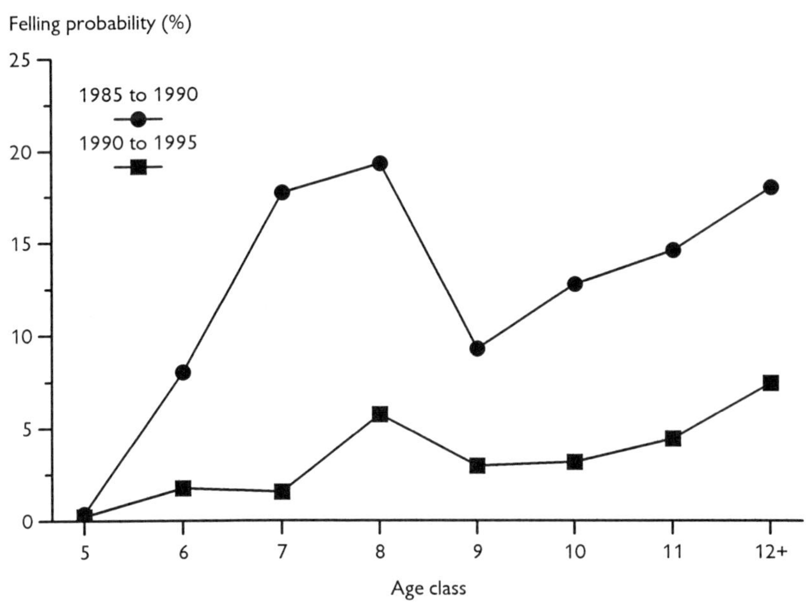

Fig. 3.6. The decline in felling probabilities

The fall in these probabilities is caused, not by changes in the characteristics of the forest owners, but by factors in the timber market and rural society. The main ones are, of course, timber prices and costs. As was seen in Fig. 1.4 in Chapter one, the nominal price of timber is now lower than in the past and, at the same time, wage inflation has increased costs and depressed profitability. As this is the case it is worth looking in a little more detail at the economics of Japanese timber production.

The economics of forestry in Japan

Timber prices

International competition is the main factor keeping timber prices low in Japan. Despite the price premium which it can command, even hinoki's price follows that of sugi which is in direct competition with imported timber. Figure 3.7 shows the path of sugi log prices and *beizai* log prices, measured at the factory gate. The former are medium-sized logs measuring 14 to 22 cm by 3.65 to 4 metres. The *beizai* logs are Western hemlock logs, sized at least 30 cm by at least 6 metres. The close correlation between sugi and *beizai* prices is clear. Also, the diagram shows that the price of *beizai* has risen relative to sugi. In the 1960s and 1970s, for example, sugi log prices were around 10-20% higher than Western hemlock logs. However, now this price premium has been lost and recently even reversed. This is consistent with the survey on the use of sugi which was reported in Chapter two. There is was suggested that the use of sugi was increasing on price grounds rather than as a result of any inherent Japanese character that the timber might have. The implication is that the increase in western-style housing reported

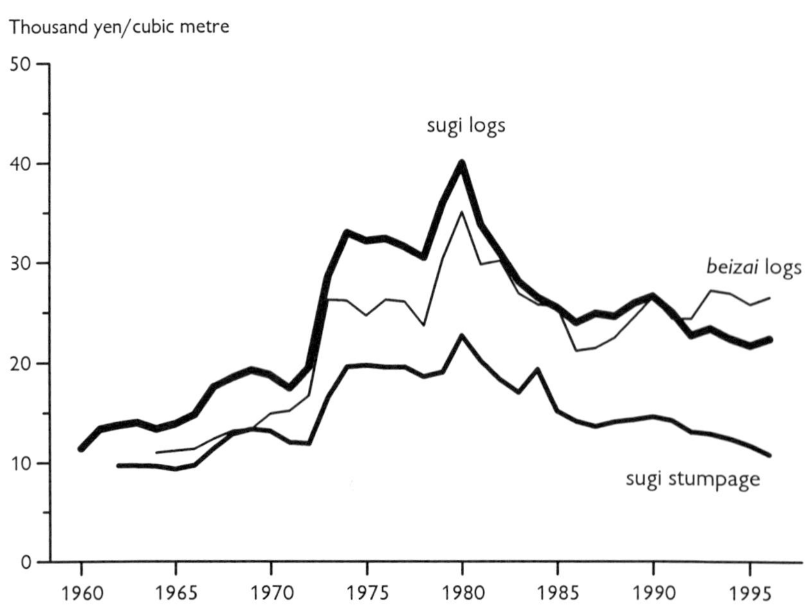

Fig. 3.7. Sugi stumpage and log prices and import prices

in Chapter two is adversely influencing the price of sugi.

Recently the price premium that hinoki logs command is also showing signs of going into decline. In the post-war years the premium increased from around one-quarter to nearly two-and-a-half times. Recently, however, it has started to fall, suggesting to some that the market for the timber is limited and that the preference for it over other species is not sufficiently strong for it to expand its market share and maintain its price premium at the same time (Kumazaki, 1998). Again, this is consistent with the discussion of the housing market in Chapter two.

Thus, the dominance of imported timber means that the prices of sugi and hinoki will be influenced mainly by costs in North America, New Zealand, Canada, Russia and the exchange rate.

Despite the trends outlined above, in the past the visible features of the wood used in traditional housing have lead to some peculiarities in the timber market which might be of relevance here. For example, the price-size gradient exhibited by sugi is shown in Fig. 3.8. The diagram, adapted from Sakaguchi (1983), shows the prices for sugi logs at various auctions.

The form of the graph is very different from that which is usually shown in, say, British forestry where the unit price increases smoothly with the size of tree. In Japan, the dominance of the market for columns is such that, when the tree is big enough to form a column of 10.5 cm or 12 cm, its price rises and this is the

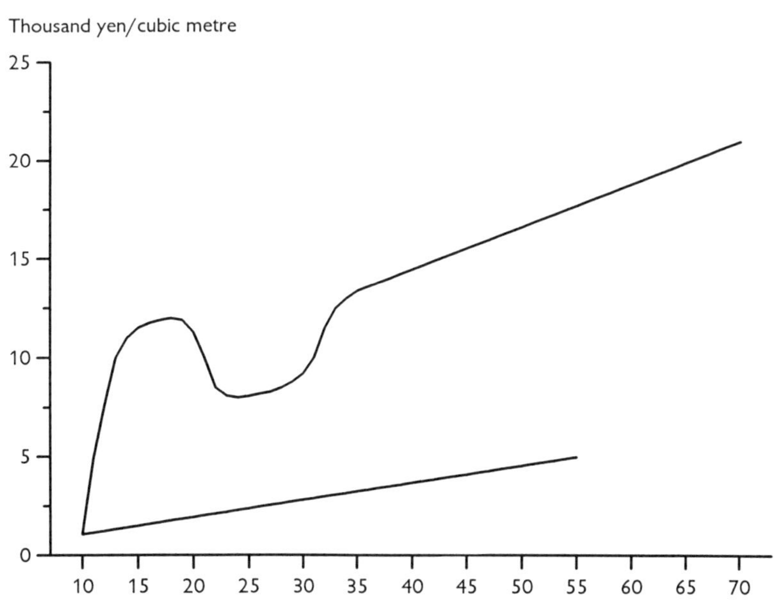

Fig. 3.8. The price-size gradient of sugi

meaning of the first peak in the diagram. After that, despite increasing in size, it is not very much more valuable, indeed its unit price falls. The best cutting pattern would still be to cut a square out of the middle and discard the other parts of the timber. However, once a much larger bole size is reached, the tree can either be cut into four pillars or else enters the market for higher value timber and so its unit price rises once more.

The lower limit in Fig. 3.8 shows the lowest prices achieved at auction and these were for logs of low quality which would only enter the market for very low grade columns or other uses. Here the peculiarities of the Japanese housing market do not intrude and so the price-size gradient is more conventional.

Whilst it would seem reasonable to assume that hinoki timber would follow the same pattern there is some doubt as to whether it actually does. Some argue that such a pattern is shown. Ishihara (1980) says that the ‘dip’ in price occurs around the 20 cm dbh mark, as it does in sugi, while Fujimori (1985) states categorically that such a pattern does not emerge. His argument is that part of the reason for the pattern exhibited by sugi arises from the fact that its heartwood is heavily coloured and this starts to be visible in pillars taken from larger-sized trees and reduces their value. Hinoki, on the other hand, shows no such colouring and so shows no such drop in value.

In *Sugi no subete* (Sakaguchi, 1983), from which Fig. 3.8 was taken, figures for sugi from the Yoshino region are also presented. This is one of the regions with a long history of timber production which were mentioned in Chapter one. It is still famous for high quality sugi. The prices for Yoshino sugi reported are far in excess of the upper line in Fig. 3.8. For example, timber in the 15 cm to 20 cm butt-end diameter typically fetched prices of around ¥150,000 a cubic metre while in the upper range of diameters prices double that of the best prices received for other sugi were common.

In the original text further data are given for the annual ring widths and degree of twist in sugi logs. The figures show that for small log sizes, i.e. those which are in the first range of diameters in Fig. 3.8, the price for logs with average annual rings of about 6 mm was only a half or a third of that commanded by logs with annual ring widths of only 2 or 3 mm. A similar trend was shown for the straightness of bole but this was somewhat less marked. In general, any degree of bend reduced the price but only by about a quarter.

Two points emerge from the discussion above. One is the rather strange fact that the Japanese have not done more research on this subject, which is clearly so important for the economics of forestry. The author is not aware of very much more work in this area other than the references cited above, all of which are old. Secondly, some argue that the pattern of prices shown in Fig. 3.8 means that the increase in output from Japanese forests will emerge, not in the near future, but when the forests have entered the age classes associated with the higher value, large diameter timber. Thus, the argument runs, owners are waiting for the second peak in Fig. 3.8 before felling. The second peak in the diagram, occurring at around 40 cm, will not be reached until the trees are around 100 years old, or more. In this case, the output of timber will enter the market when the peak of the area distribution shown in Fig. 1.7 reaches age class 20, a delay of about 50 years.

Wage costs and productivity

Figure 3.7 also shows the stumpage price of sugi and this has been even more depressed than the log price by foreign competition, and the same is true of hinoki. It is in the final product market where competition actually occurs and so it is the cost of lumber which will be equated by competitive pressures. This means that the stumpage price of sugi must allow for the extraction and processing costs to be covered and, given rising costs and a relative inability to increase productivity, this gap in prices will increase.

Having said that, recently there has been a significant increase in the amount of harvesting technology in use in Japanese forestry and this could go some way to increasing the productivity of labour. For example, the number of feller-bunchers in use rose from only four in 1988 to 50 in 1996. The number of processors rose from seven to 596 in the same period (*Rinya jihou*, 1/97; Forestry White Paper, 1997). Figures show that most of these machines are in use in Hokkaido, Touhoku and Kyushu. However, the nature of the topography in Japan is such that the wide-spread use of such machines is limited. The regions where they are now in use have, if anything, slightly more favourable topography than the rest of Japan but the potential to use machines is, at best, limited. Under favourable circumstances this number of processors could have a significant impact but, apparently, only 5% of the felling gangs operate such machines and the down time is around 50%.

At the same time, of course, wages have been rising. The overall wage index has risen nearly four times in real terms since 1960. This, coupled with the drop in real prices and difficulties in increasing productivity, has drastically reduced profitability.

The internal rate of return of Japanese forestry

One of the statistics that the Japanese government publications often contain regarding the economics of Japanese forestry is the internal rate of return (IRR) of plantations, and their results show the extent of the problem caused by these adverse economic movements.

The IRR is usually calculated in the Forestry White Papers by assuming a simple point-input/point-output model. An area is assumed to be planted, at a cost dependent on the prevailing costs of labour, seedlings, etc., and the IRR is calculated with an assumed rotation length and standing timber sales. The results of such calculations are shown in Fig. 3.9 and are adapted from the Forestry White Paper for 1992. Despite the simple model underlying the calculation, the trend is quite clear and it is the obvious one. An interesting element in the graph

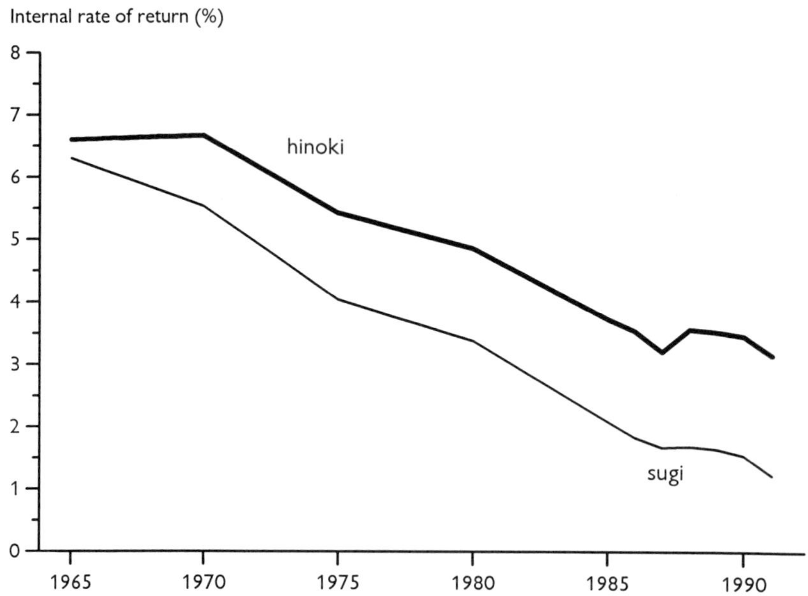

Fig. 3.9. The internal rate of return of sugi and hinoki

is the fact that, of the two species highlighted, the IRR of hinoki is declining less rapidly than that of sugi. But as Fig. 3.1 showed, it is the output of sugi which is showing the upturn whilst that of hinoki is continuing to decline.

The IRR in Fig. 3.9 is, of course, an average figure and some forests will be more profitable and some less. Oka (1997) argues that Japan's forests can now be divided into four types. First are those with positive net present values. Second are those capable of generating a positive cash flow on felling in excess of the regeneration costs but not enough, when capitalized over the rotation, to give a profitable investment. These will have negative net present values. Third are those with a positive stumpage value but not of a sufficient magnitude to cover the costs of regeneration. If felled the area would be left unplanted. Finally are those with zero or negative stumpage values. Sadly he goes on to state that it is difficult to estimate the relative proportions of these types of forests. However, the forecasts of MAFF contain a similar classification and, according to their figures, only 40% of forests were economic at the beginning of the 1990s suggesting that nearly half of Japan's forests are in the last two, or possibly three, groups.

Oka also gives examples of the range of possible costs and returns to Japanese forestry taken from the latest stumpage price and cost surveys carried out by the Forestry Agency and these are adapted here in Table 3.7. The figures show the very low returns from standard Japanese forestry practice although the low rates of return and low inflation rates of Japan should be kept in mind when

Table 3.7. The profitability of coniferous species

		Sugi	Hinoki	Larch
Price at sawmill		22,200	53,500	12,900
Logging cost	Maximum	23,650	41,050	9,620
	Minimum	1,175	2,056	2,949
Transport cost	Maximum	6,000	4,465	3,626
	Minimum	926	1,431	1,298
Average cost		11,041	14,188	8,277
Regeneration cost		2,200,000	2,400,000	1,000,000
IRR	Maximum	2.2%	3.9%	0.8%
	Minimum	-1.2%	0.0%	-0.9%
	Average	0.5%	2.2%	-0.1%

Price, logging and transport costs - yen/cubic metre; regeneration costs - yen/hectare.
Yields: sugi & hinoki 300 cubic metres/hectare, larch 200 cubic metres/hectare.
Rotation length - 40 years.
source: adapted from Oka (1997)

making international comparisons. Of course, the climatic conditions of Japan lead to relatively high regeneration costs. Much weeding and climber cutting is needed to keep the trees free from early competition and, in the more intensive management systems, snow is cleared from the young trees to prevent breakage.

Two responses have been suggested to combat these low IRRs. One is a reversion to natural forests and the second is the adoption of longer rotations. A move to natural forests would be welcome by many Japanese as the pollen from sugi in man-made forests causes epidemics of hay fever. The trend seems to have already begun. The area of natural forests is on the increase. Although the overall forest area figures do not seem to reflect this trend, the figures for the areas felled and replanted do. Until around 1975 the reported planted area was running at around 75% to 80% of the area reported as felled. Since then there has been a continuous decline in the proportion replanted and now the figure stands at around 25%.

Whilst this will have little direct effect in the short run, in the long run one might expect a gradual change in the species composition of the forests. Unless careful systems of felling are undertaken, it is quite likely that logged over areas will revert to the species composition described in Chapter one in the section on natural forest types, the point being that most of these species are less valuable and not of direct use in the important area of building.

The decline in the planted areas of the various species is also informative. Overall the area planted in 1992 was only 13% of the 1960 figure. The figure for sugi was 11%, that for karamatsu 5% and for other conifers the figure was 6%. Only hinoki has shown any resistance to the trend, with the area planted falling to 'only' 29% of the 1960 level.

The argument that rotations are being intentionally lengthened receives some support from estimates of the management of hinoki and sugi which show higher rates of return compared with those in Table 3.7. This is partly the result of the increase in the unit price shown in Fig. 3.8. For example, Iehara (1993) gives an internal rate of return of over 4% for a rotation length of 100 years. However, Blandon (1991a) found that the increase in unit price as the timber moves into the second peak in the price-size gradient was generally not sufficient to compensate for the extra wait for harvests and the highest profitabilities were to be had at standard rotations of around 40 and 50 years.

The decline in the IRR of the coniferous species is, if anything, somewhat understated by Fig. 3.9. This is especially the case if a longer term perspective than that shown in the diagram is taken. In the early post-war years, not only were prices high but the level of technology was such that small-scale holdings were not so relatively disadvantaged. Also, as was explained in Chapter one, the labour used was often provided by family members during the agricultural off-season. Given the absence of other paid employment, the opportunity cost of such labour would have been less than that implied by the statistics for labour costs. However, with the growth of the economy, the availability of paid employment in civil engineering works, local factories subcontracting for large firms, etc., has expanded greatly such that the opportunity cost of forestry work has come into line with the reported wage figures. Thus, the increase in the true costs of forestry will, from the point of view of private forest owners, have been greater than that implied in calculations such as those represented by Fig. 3.9.

This trend is picked up in the Forestry Censuses. Some results are given in Table 3.8 which shows the main employment of forest households having a forest holding of 1 hectare or more (5 hectares or more for the 1960 figures). In 1960 a typical agricultural/forest household would have been self-employed and engaged in agriculture. Some would have been mainly forest workers but these would have been the households with the largest holdings. Now, the typical picture of such a household is one that works full-time for another person or company. As would be expected, the latest census shows that the number of households fully employed outside of forestry and agriculture decreases with an increasing size of forest holding. But even for the largest size of holding of 100 hectares or more, only around one-fifth reported their main occupation to be forestry.

The trends shown in Table 3.8 will also be associated with the average age of

Table 3.8. Employment of forest households

	Employed by others		Self-employed		
	Full-time	Part-time	Forestry	Agriculture	Other
1960	5.1	3.0	4.5	78.1	9.2
1970	19.5	9.1	1.8	58.0	11.6
1980	36.2	10.3	1.2	37.5	14.7
1990	46.9	6.4	1.0	26.5	19.2

All figures are percentages.
source: Akaha (1992)

labour in forestry and reported in Chapter one. It is usually the younger members of the family who can find full-time employment outside of the rural areas.

Thinnings

One bright point in Japanese forestry is the recent increase in the level of thinning activity. The close planting densities for sugi and hinoki reported in Chapter one were worrying the Japanese for, without thinning, the quality of the resulting timber would be much reduced. So to increase the amount of thinning undertaken subsidies were introduced in 1981 and resulted, not surprisingly, in a sudden increase in the level of thinning output.

Recent figures from the 1997 Forestry White Paper show that around 4 million cubic metres of thinnings are now being taken out each year. Around half of this volume is non-economic and does not reach the market. Of the remainder most is used for lumber in building or for packaging use. This is an area where co-ops dominate and the policy to expand thinnings has largely been carried out by them.

However, the size of the timber means that its use is limited and it is unlikely to be able to compete with imported raw material. Also, the nature of the topography of Japan means that thinning activity is likely to be relatively even more expensive than it is in other countries.

The Japanese Forecasting Method

The forecasting method used by the Japanese has already been alluded to and, although a discussion of the forecasts of domestic output will be left until Chapters seven and eight, it is useful here to mention briefly the way in which the Japanese forecasters have reacted to the changes outlined above.

The forecasts of output of coniferous timber are formed by multiplying the felling probabilities such as those shown in Figs 3.5 and 3.6 by the area of the age class in question. However, the Japanese method does not see the felling probabilities as being fundamental. Rather they are seen as being derived from an underlying probability distribution referred to in Japanese as the *gentan* probability distribution, where *gentan* is a Japanese word meaning reduction in area. A brief aside on the statistics of the method might be useful here.

Gentan probability theory

Gentan probability theory was developed by Professor Suzuki of Nagoya University and was originally used as a means of predicting or planning the output of private forests in a particular area, especially in the early years when attempts were being made to reduce over-cutting by private forest owners (Suzuki, 1972; 1973; 1979).

According to the theory, because of the multitude of small forest owners in Japan all felling for different reasons, the felling of a stand can be seen as a random process. Suzuki argued that this process could be modelled using simple life data analysis. The forest was seen as a product which would 'fail', i.e. be felled, according to some distribution. This distribution describes the gentan probability, the probability that a forest planted now would be felled in a certain time period in the future. From this gentan probability distribution a set of conditional probabilities can be derived showing the probability that a forest which has survived until now will be felled. These are the theoretical equivalents of the felling probabilities which were illustrated in Figs 3.5 and 3.6.

This approach was modified by a concept known as the economic forest ratio. The argument ran that, because of the fall in the real price of timber some forests had become uneconomic in the sense of the categories of forest developed by Oka (1997) and described above. The ratio of economic forests to uneconomic ones was found to depend upon the price of timber and the length of forest roads built. In the mid-1970s, this economic forest ratio was estimated to be in the region of about 50% but it seemed to fall off quickly in the early 1980s to reach levels of around a third. The forests in the uneconomic group were excluded from the gentan probability modelling process. Output from these forests was forecast using a low felling probability.

The way in which the theory was used can be summarized as follows. Work by Amano and others for the 1987 forecasting exercise indicated that the mean and variance of the felling age were not stable through time but that they seemed to be increasing (Noda and Amano, 1983). This is consistent with the drop in felling probabilities seen in Fig. 3.6 and with the discussion on the economics of forestry. Further work by Amano and Noda (undated) and Amano *et al.* (1984) can be summarized as:

> '.... the mean rotation age of 50 years in 1972 increases gradually to over 55 years at present. This increase, however, is linear and would not [present] any problem for projecting the mean trend into the future. ...
>
> The situation is not so simple with the variance of rotation age, [which] since the oil crisis has remained rather stable in the roadside forest ... but [has] fluctuated widely with time in remote forests. ... In other words, the harvesting and subsequent regeneration have been carried out regularly in the roadside forests, whereas the operations have become more and more erratic in the remote forests. This fact suggests that the gentan probability for projecting timber supply from the remote forests is inappropriate.' (Amano *et al.*, 1984)

Thus, for the purposes of forecasting, all of the discussion above can be condensed into the mean and variance of the gentan probability distribution and the economic forest ratio. The upshot was that all of the adverse movements in the timber industry can be expressed as changes in the economic forest ratio and the parameters of the gentan probability distribution. In fact, it was assumed in the earlier forecasts that the movements in the felling ages had come to an end and the mean and variance of the gentan probability distribution would either remain stable of increase slowly. Therefore, the fall in the felling probabilities shown in Fig. 3.6 was attributed to more and more forests becoming uneconomic. The felling probability for those forests in the economic group was thought to remain more-or-less unaltered. Sadly, the forecasts did not relate the parameters of the gentan probability distribution to the characteristics of owners in the way which North American and Scandinavian research suggests would be possible.

Natural Forests

Much of what has been said concerning the economics of man-made forests applies equally to natural forests and so will not be repeated here. As is usual for hardwoods, there is a premium for quality but the criteria are more widely defined than they are for conifers (Tsukamoto and Fujiwara, 1995).

The age class distribution of natural forests shown in Fig. 1.8 in Chapter one indicated that most of the national forests were located in the older age classes and so the idea that there was a large overhang of volume approaching maturity did not apply. On the other hand, the area of natural forests in private hands did show an age class structure similar to that of the mainly coniferous man-made forests. It might be expected, therefore, that a very similar type of theory to that just described could be used in the forecasting of output. Unfortunately, however, the data which are collected concerning broad-leaved forests are not as detailed as they are for coniferous ones and so the degree of 'resolution' that can be obtained in deriving figures for felling probabilities is much less.

That said, the Japanese forecasts, as will be seen in Chapter seven, do not apply such a method but rather utilize a simple trend extrapolation technique. To the extent that the mainly hardwood forests no longer really figure in the supply of sawnwood to the Japanese economy this represents no real problem as the main area of interest lies in the possible expansion of output in the lumber area. Thus, it seems that the Japanese do not really consider the natural forests as being as interesting or significant as the man-made coniferous ones. Indeed, as Table 1.5 showed, the broad-leaved species mostly found in natural forests go mainly for paper production and in this area the overall self-sufficiency ratio is only 12.9%. Also, the market is not one where there are any peculiarities such as those dictated by parts of the housing market and the use of hardwood timber for sawnwood has been declining very rapidly.

So it is probably true that any increases or decreases in output from natural forests could be forecast using the fairly straightforward extrapolation techniques which the Japanese use.

Conclusions

This brief look at the Japanese forestry sector has shown a part of the economy more-or-less open to foreign trade. As a result it is suffering reduced prices and depressed domestic production as more efficient foreign producers are able to undercut the higher priced, less efficiently produced domestic product. This is a somewhat different story from the one usually associated with post-war Japan.

Given the difficult terrain and environmental uses of Japan's forests, along with the likelihood of the increasing lack of interest on the part of forest owners for timber production, it is tempting to forecast the output of Japan's forests by merely extrapolating the downward level of output into the future. However, the method adopted by the Japanese for forecasting shows the problems of doing so. Given the age class structure of Japan's forests the output of coniferous timber has the potential to increase dramatically.

Thus, if nothing changes it is possible to extrapolate a falling output or predict an increasing one. The reason for this apparently self-contradictory statement hinges on what is meant by 'nothing changing'. If it is argued that the felling probability for a given age class reflects the behaviour of forest owners, as it is determined by the interaction of economic and social factors described here, then the idea of nothing changing implies that these probabilities will remain constant into the future. If nothing alters the economic forest ratio, the output of forests in Japan will increase in proportion to the increase in areas entering the age classes when most felling occurs. Thus, the notion of nothing changing is actually compatible with a large increase in domestic output and, therefore, reduced imports.

On the other hand if the phrase 'nothing changing' implies that the trend of changes in behaviour will continue as before it follows that the fall in the felling probabilities such as that shown in Fig. 3.6 will continue either by the economic forest group dropping in size or else by the mean and variance of the gentan probability distribution altering. Depending on the rate of fall of the probabilities the output from Japan's forests will either increase or decrease.

It would seem feasible, although the data do not seem to support this, that most of the trends acting to reduce felling probabilities and the size of the economic forest group are played out. The movement of population from rural to urban settings has proceeded apace since the war but now Japan's population distribution is similar to that of most other industrialized nations. The ageing of the rural population is likely to level off and the drop in real timber prices has also probably come to an end. Thus, it is very tempting to argue that the drop in felling probabilities, if they represent the result of the interaction of these factors, is also coming to an end.

It is also tempting to argue that there could even be a reversal of the trends which have been seen in the past. As a broad generalization it is true to say that the Japanese have, over the post-war years, seen urbanization and moving from the countryside to the town as an unambiguously good thing. Environmental protests, campaigns against road and rail construction which have been seen in other countries are only really just beginning in Japan. It is true to say that, for the most part, the rural areas are depressed in relation to the town. However, anecdotal evidence of a reversal of this trend has emerged in recent years with businessmen indulging in hobby-farming, school-leavers claiming to opt for universities and colleges in more amenable surroundings of small towns, etc. Narita (1996) presents a case study of forest management in central Japan where a number of workers are returning to forestry because they are 'tired of work which presses for quotas under time pressure'. He concludes that the firm under study is 'evolving into a company of young people quite unlike typical forest companies in Japan'.

If this trend is actually occurring it could signal a beginning of a resurgence of the rural economy but, if the example of Britain is anything to go by, this would mean not the resurgence of agriculture and forestry but rather an increase in the use of forests for environmental and recreational uses. Thus, any move to rescue the forests from their present situation of semi-abandonment would not necessarily lead to any increase in the output of timber.

All this means that the forecasting of output from Japan's forests is a very difficult operation. Slight changes in the ideas of what is happening in the rural economy lead to large changes in the level of timber output making the forecasts more sensitive than most. However, such problems will be left now until Chapters seven and eight when the results of the Japanese forecasters' and the author's models are discussed.

Another factor, not so far examined, is the fact that the foreign suppliers of timber are also undergoing changes in their circumstances. If these changes are to be reflected in an increase in import prices or species composition it might be the case that they could trigger the increase in domestic supply. So, it is time now to look at the imported supply of timber products.

Chapter four:

Imported Timber Supply

Introduction

Over the years after the Second World War Japan has become more and more dependent upon imports for its timber requirements. Recalling the figures from Chapter one, the overall self-sufficiency ratio has fallen steadily since imports started and now stands at around 20%. The dependence upon imports for plywood and veneer is the most marked and, indeed, it was in this sector that imports of logs began. However, now the sawnwood sector is only around one-third self-sufficient. This is a more important figure as it is in this sector of the timber market that the largest part of Japan's burgeoning growing stock would most suitably be targeted. This high dependence on international markets is, of course, welcome news for exporters of timber as, given the peculiarities of the timber market there, Japan is often willing to pay a premium for good quality timber in its construction industry.

The question posed in this book, and in Japan, is whether this dependence on imports will continue or whether an 'era of domestic supply' or *kokusanzaijidai* is just beginning. The upturn in the output of sugi shown in Fig. 3.1 in the previous chapter is causing a little interest and suggests, to some, that such an era is at hand. The rapidly maturing growing stock suggests that the biological potential for an increasing supply is there but, as the previous chapters showed, the economics of forestry and the trends in the markets make any large increases in supply seem unlikely.

Should such an increase in output occur it would most probably appear in connection with the more valuable species of sugi and hinoki as these remain the most profitable to harvest. As these species are mostly used in the construction of housing, the possibility of a *kokusanzaijidai* is of more significance to some exporters of timber than to others. Most notably, the exporters from North America would seem to have most to lose if Japan's domestic timber industry realizes its full potential as their timber is in more direct competition than that of other exporters. However, given the recent trend towards the use of coniferous plywood and other panel products as well as the ability to substitute these for solid timber in housing, it is possible that the repercussions of an increase in domestic coniferous supply could reduce the demand for these products, either by people moving back to solid timber or by the use of domestic species in the panel products market.

As one of the main determinants of domestic supply is the price of timber, and as this price is largely determined by imports, a major factor is the exchange rate. This is a variable which, given developments at the time of writing, is even more difficult to predict than usual and is, anyway, outside of the scope of this book. Therefore, factors affecting the exchange rate, important though they are, will not be considered.

Any increase or decrease in the harvest levels of the exporting countries will have an affect on the Japanese supply potential by reducing or increasing the price of timber. A number of the producing countries are facing problems in maintaining their export position, with environmental and other pressures causing harvest reductions. Some, such as New Zealand and Chile, are looking to increase timber output and will be attempting to expand their exports to Japan. It is against this backdrop that any forecasting exercise has to be carried out. In the next two chapters the situation in the individual exporting countries will be examined but, here, the overall import position will be examined.

The Overall Import Position

The overall position for timber imports has been outlined in Chapter one, where figures were presented which showed that the self-sufficiency ratio had dropped to 20% for timber products in total. Recently, MAFF has started to produce data on the imports of timber products into Japan broken down by country of origin and in RWE. The 1996 data are given in Table 4.1 and refer to all end-uses and correspond to a total demand for timber in the country for that year of 112,324 thousand cubic metres RWE. As was noted in Chapter two, the other data which the Japanese and other agencies publish are in unconverted units. Thus, the figures in Table 4.1 will not necessarily correspond to other published data.

Table 4.1 shows the predominance of timber from the North American continent which constitutes a greater amount than the Japanese contribution to timber demand. The demand for *nanyouzai* or tropical timber is running at about 14% of total timber demand and the dependence on Southeast Asia is clear with very little timber, as yet, coming from Africa or Latin America.

The Japanese publish a consistent set of these figures back to 1991 and the proportion of the imports from various countries in that year is also shown in the table. The overall level of demand in 1991 was not much different, standing at 112,202 thousand cubic metres RWE. However, one or two interesting trends

Table 4.1. The sources of timber and timber products

Country of origin	Imports for 1996 (000 cubic metres RWE)	Share in 1996 (%)	Share in 1991 (%)
USA	27,961	24.9	28.0
Canada	14,678	13.1	10.6
Beizai total	42,639	38.0	38.6
Malaysia	7,420	6.6	10.1
Indonesia	6,688	6.0	4.9
Papua New Guinea	2,010	1.8	0.8
Other tropical timber	497	0.4	0.1
Nanyouzai total	16,614	14.8	16.3
Russia (*Hokuyouzai*)	5,722	5.1	4.4
Chile	4,240	3.8	4.3
Australia	6,118	5.4	4.5
New Zealand	3,546	3.2	3.2
Other	10,961	9.8	3.7
Japan	22,483	20.0	25.0
Total	112,324		

Some figures contain rounding errors.
source: Forestry White Paper (1996)

are visible even over this short period. Of course, the drop in the self-sufficiency ratio is clear with the Japanese sources of supply falling. The level of imports from North America has remained more-or-less constant but there has been a shift in favour of Canada reflected in a movement to sawnwood imports and away from log.

The level of tropical timber imports has dropped slightly but the level of Malaysian imports has shown the greater fall. The increase in imports from Indonesia mainly reflects the growth of plywood imports shown in Fig. 2.9 in Chapter two. The Japanese are beginning, slowly, to obtain tropical timber from elsewhere with imports from Papua New Guinea and other countries increasing. The category 'other' shows a marked increase, from 3.7% to 9.8% indicating a general diversification of timber sources. The increase in the proportion accounted for by imports from Russia is notable given the problems in that country over the years in question.

Some of the countries shown in Table 4.1 will be discussed in the next two chapters. First, however, it is useful to examine different types of import and look at the trends that have occurred over time. Here, the breakdown will take the form of logs (for all end-uses), lumber, plywood and panel products and pulp. The overall self-sufficiency ratio of 20% varies across these end-uses. The corresponding figures for the four main end-uses which the Japanese identify in their figures were 32.5% for logs and lumber, 12.9% for pulp and chips, 1.4% for

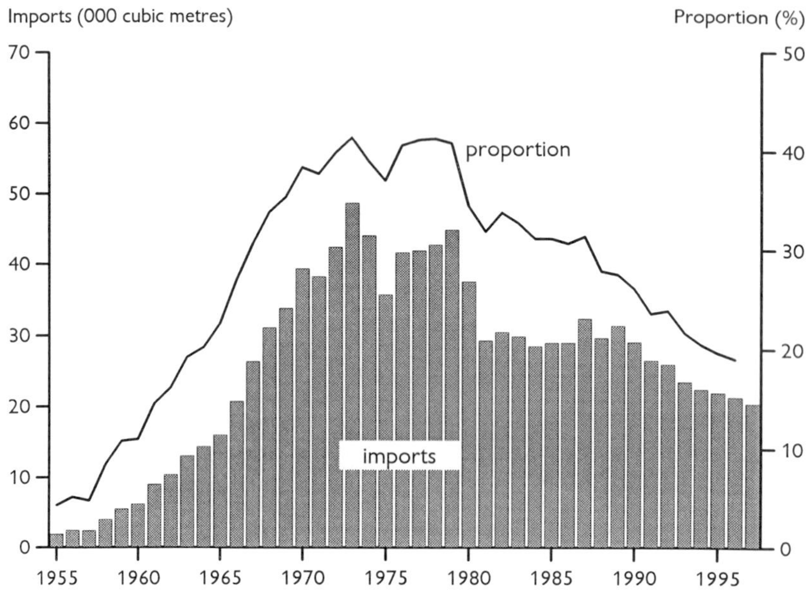

Fig. 4.1. Log imports

plywood and veneer and 14.2% for other uses (Forestry White Paper, 1997).

Log imports

Figure 4.1 shows the level of log imports from all countries for the period from the 1955. The figures account for all types of log, from saw logs to those for pulp and veneer use and are taken from Ministry of Finance data. The familiar pattern emerges with demand for logs increasing rapidly until the oil shock in 1973. After that log imports started to fall against a background of a fairly stable level of total timber demand. This is part of the trend towards the import of finished and semi-finished goods. One aspect of this is shown by the part of Fig. 4.1 labelled 'proportion' which shows the proportion of total timber demand, in RWE, made up of log imports. This falls in line with the drop in log imports after the oil shock of 1973 reflecting the roughly constant total demand for timber products.

The data in the diagram contain two main categories of importance here, that is coniferous and non-coniferous species and the composition of this trade has changed over the years. In the early 1960s the proportion of coniferous logs was only around one-third of the total as imports of hardwoods for the plywood industry were given precedence. However, after that period the level of coniferous imports began to rise and now stands at nearly 70% of total log imports. The sources of log imports in 1997 are shown in Table 4.2 along with

Table 4.2. The sources of logs for the Japanese market

Source	Imports for 1997 (000 cubic metres)	Share in 1997 (%)	Share in 1977 (%)
USA	5,638	27.6	
Canada	118	0.6	
Beizai total	5,756	28.2	25.3
Malaysia	2,974	14.6	22.8
Papua New Guinea	1,945	9.5	
Other tropical timber	401	2.0	
Indonesia			22.9
Philippines			4.0
Nayouzai total	5,320	26.1	51.1
Russia	6,137	30.1	20.8
New Zealand	1,983	9.7	2.1
Europe	155	0.8	
Africa	662	3.2	
Other	394	1.9	6.0
Total	20,407		

source: Forestry White Paper (1981, 1997)

the proportions for the total log imports in that year and 20 years earlier in 1977.

Comparing Tables 4.1 and 4.2 shows the reduction in the importance of Canada, which has long operated a log export ban from Crown forests in British Columbia, and Indonesia which introduced a total log export ban in 1985 and then replaced it in 1991 by prohibitively high export tariffs on logs. Australia also disappears, exporting mainly chips to Japan. The figures for the proportion of log imports in 1977 show the way in which Japan's sources of supply have slowly diversified. Also, the source of tropical timber differs with Indonesia and the Philippines appearing in the earlier data as log sources but later not warranting a separate entry.

The log imports in 1977 were 41,880 thousand cubic metres, almost double the most recent figure. Against this, the proportion for imports from North America show almost no change. Similarly, the high proportion of *nanyouzai* imports shows that the importance of Japan for the world log trade was more significant in the past than it is now. FAO figures similarly show that Japan's share in the world log trade was over half in the 1960s and 1970s but now has dropped to only around one-quarter.

Coniferous logs

The imports from North America, Europe and New Zealand will mostly be in the form of coniferous species and, it is in this area that Japan exerts a very large influence on world trade. The present position of Japan in the world markets is

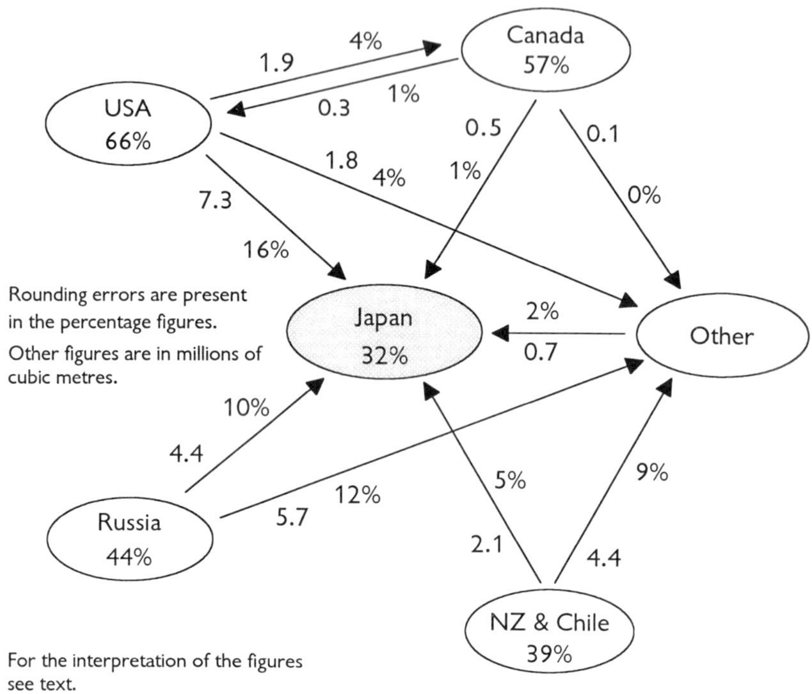

Fig. 4.2. The position of Japan in the coniferous log trade

shown in Fig. 4.2. This diagram is prepared using FAO data for 1994 and summarizes their trade flow data for the category 'coniferous industrial roundwood/wood in the rough' and so the figures differ slightly from those found in the Japanese data.

The diagram indicates that Japan accounted for 32% of all coniferous roundwood traded on the international markets. Considering the USA, the trade flow from the USA to Japan amounted to 7.3 million cubic metres and this represented 16% of total coniferous log flow. From the point of view of the USA, the Japanese market absorbed 66% of total coniferous log exports. The other figures in the diagram are to be interpreted in the same way. The diagram shows not only the dominant position which Japan has in the world market but also the role it plays in the export trade of the supplying nations. However, the other countries have also been diversifying their markets and so not only has the importance of Japan in the coniferous trade dropped somewhat, so too has the significance of Japan for the various countries. Twenty years before, in 1974, Japan accounted for 79% of the coniferous log exports from the USA, 75% of those from the USSR and 98% of those from New Zealand.

The diagram shows that 25% of world trade is accounted for by exports from the countries noted to the area denoted as 'other' and the remaining trade flows, which are not identified in the diagram, are mainly the result of trade between countries within this 'other' group. Of significance here are China and Korea. The latter is the second biggest importer of coniferous logs, accounting for just under 13% of world trade in 1994 but it competes directly with Japan for imports from the USA, Russia, New Zealand and Chile. Indeed, as far as the last two countries are concerned, Korea is the bigger customer. China, whilst accounting for far less of the world market, is important in that, like Korea, it obtains a lot of its coniferous log supply from the same sources as Japan.

Thus, growth in these two countries will have a direct influence on the ability of Japan to continue to import coniferous logs. As yet, however, Japan tends to import higher quality logs than these two countries but if the trends in the housing market discussed in Chapter two continue, this will not be a position which will be maintained in the future. Thus, for log prices, and therefore domestic log supply in Japan, the rates of growth of demand in China and Korea are likely to be significant.

Hardwood log imports

The position of Japan in the hardwood log trade has also been one of dominance. The most significant role which Japan plays is in the tropical timber trade, with its position in the smaller non-tropical hardwood log trade much less significant in both absolute and relative terms. Until the late-1980s Japan accounted for over half of the all of the tropical hardwood logs traded in the world. However, at the end of that decade the role of Japan rapidly declined and now it accounts for 'only' 26% of the total hardwood log trade. According to 1994 FAO data Japan's absorbed 39% of the tropical hardwood logs traded on the world market.

The overall position of Japan in 1994 is shown in Fig. 4.3 which is, again, adapted from the FAO directions of trade data. The dominance of Japan is apparent as it was in Fig. 4.2 but this diagram shows that, in this market, the position is not so unambiguous. The other Asian countries of Korea and China are competing with Japan and the source of supply for all of these countries is not well diversified.

Part of the reason for the drop in the proportion of the tropical log import trade relates to Fig. 2.9 in Chapter two which showed the decline in domestic plywood production and the rise of plywood imports. With this trend the demand for tropical logs decreased. However, Japanese figures for 1996 show that 83% of tropical log imports were used in plywood with 15% going for sawnwood and the rest for chips. The figures represent the end of a long decline in the use of tropical logs for lumber. In 1965 a little under half of all *nanyouzai* was used for sawnwood with all of the remainder being used in plywood manufacture.

Like the softwood log trade, Japan is in direct competition with Korea and China. These two countries are also significant in that they obtain most of their

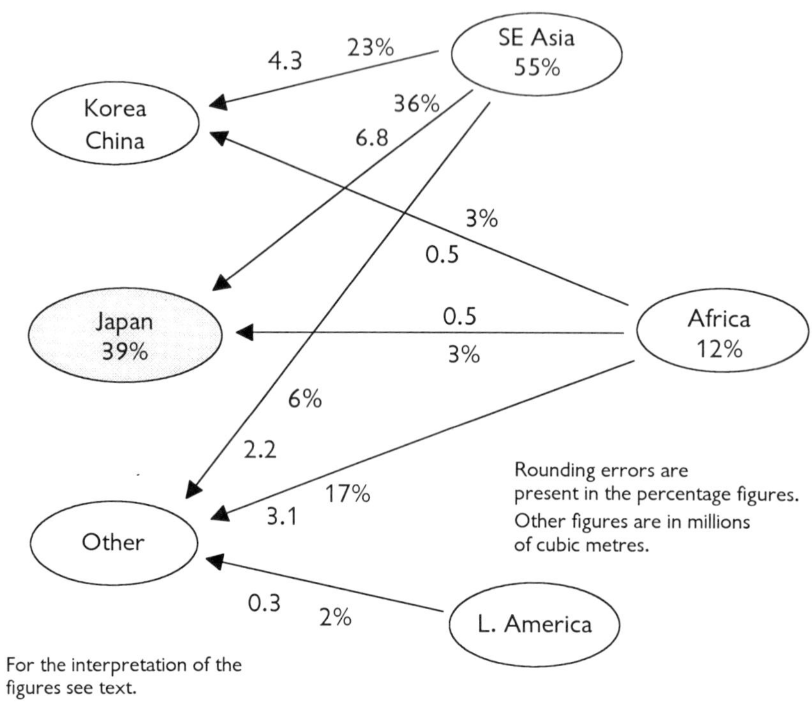

Fig. 4.3. The position of Japan in the tropical hardwood log trade

supplies from the same areas as Japan, that is Southeast Asia, although there is a small trade flow from Latin America to China and Korea. So, as was the case with coniferous logs, any increase in the growth rates of these countries will influence the ability of the Japanese to maintain their levels of imports.

Sawnwood imports

Mirroring, to a certain extent, the drop in log imports shown in Fig. 4.1 is the increase in the imports of sawnwood. Over the past decade or so the imports of sawnwood, finished and semi-finished timber products have increased. This is partly the result of policies of the exporting countries and partly a move by the Japanese to allay criticism of their policies of importing raw materials. Also, straightforward economic factors such as the rise in Japanese wages and the value of the yen, making imports of finished and semi-finished products relatively more economic, have been important. Japanese data show that the proportion of sawnwood imports has increased more-or-less continuously, now reaching a figure of around 35% of total log and lumber imports in RWE.

From the world point of view the growing significance of Japan in the

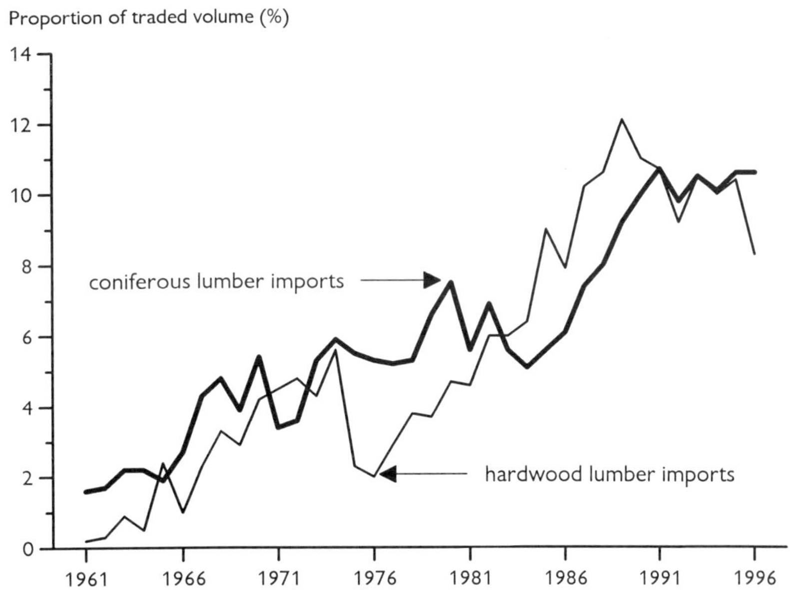

Fig. 4.4. The position of Japan in the lumber trade

markets for sawnwood is shown in Fig. 4.4 which, again using FAO figures, gives the proportion of total traded sawnwood volumes accounted for by Japan. The early years were characterized by insignificant volumes but now Japan is beginning to attain a position in this trade comparable to that in the log market.

The largest suppliers of coniferous lumber to Japan are the USA and Canada. The USA supplies one-fifth of the Japanese coniferous sawnwood market and Canadian imports account for 59%. From the exporters' point of view, Japan takes 12% of Canadian coniferous lumber exports and 39% of all such exports from the USA. However, if the Canada-USA trade is removed from the data these percentages become 71% and 51% respectively, again showing the significance of the Japan-North American trade flows. Russia, although important in the log trade, is not really significant in the coniferous sawnwood trade, accounting for only 4% of Japan's imports. There is little doubt that the Russians would like to boost their sawnwood exports to Japan in order to increase their foreign exchange earnings but quality considerations preclude this.

The tropical sawnwood trade is a lot smaller than the coniferous one and Japan is less significant in this market. About 42% of its hardwood lumber imports come from Malaysia, and Indonesia also figures.

It is probably true to say that, all other things being equal, the preference in Japan would be to import raw materials and process them within the country. It was to encourage this that the policies of 1961, the Measures to Stabilize Timber

Prices, encouraged the setting up of processing capacity around the coast of the country. The discussion in Chapter one about increasing wage levels and declining numbers of sawmills suggests that the trends in Fig. 4.4 will continue into the future and sawnwood imports will increase. As Chapter two showed, most of the lumber is likely to end up in the housing market and so the increases here will be dependent on foreign producers either being able to produce the types of product which fit in with the traditional post-and-beam style of housing or else the success of the alternative housing methods.

Plywood and panel product imports

The figures and diagrams above all suggest the continuing dominance of the import of raw materials rather than finished products but also show a movement towards the latter. This has already been seen in the discussion in Chapter two on the Japanese plywood market which has been characterized by a move away from the imports of logs to the import of finished products, to the detriment of the profitability of Japan's domestic plywood manufacturers. Also, it was seen there that the other panel products were of relatively smaller importance.

Plywood

Most of the log imports from Malaysia and other tropical logs shown in Fig. 4.3 were destined for plywood manufacture. Recently, small quantities of the logs from other countries have also been used for plywood and, of course, there has been an increase in the use of coniferous species in the area.

However, as Fig. 2.9 in Chapter two showed, in recent years there has been an increase in the importation of plywood itself. This has mostly been the result of a substitution of log exports for exports of processed products, especially in Indonesia. The source of plywood imports into Japan expressed in cubic metre RWE show that Indonesia and Malaysia account for 58% and 28% respectively of the total (Araya, 1998). Canada and the USA account for most of the remainder supplying 6% and 2% respectively. North American plywood is almost entirely coniferous, unlike the Southeast Asian countries which export hardwood plywood. Other countries account for the remainder of the trade.

Again, China and Korea are major competitors with Japan, although specifications no doubt differ somewhat. For the Indonesians, Japan takes 34% of their plywood exports according to FAO figures while China accounts for 19% and Korea 14%. For the Malaysians, China takes nearly twice as much plywood as Japan. So, once again, growth in these markets will impinge on Japan.

To the extent that the plywood industry in countries such as Malaysia and Indonesia are less efficient than those in Japan, the move to import plywood rather than logs is likely to be associated with reduced conversion ratios and increased forest degradation in Southeast Asia. The size of the cores left after logs have been peeled is a lot larger in Indonesian mills than it is in Japanese ones. Also, there is probably more waste in Indonesian and Malaysian mills as the raw material is nearer and, therefore, relatively cheaper. Indeed, there has been discussion of whether it is better for countries such as Indonesia and Peninsular Malaysia to maintain log export bans when they could make higher returns by the export of logs rather than processed products.

Other panel products

As has been noted, until recently Japan was not a large consumer of panels products. Recently, however, as part of the diversification of its sources of timber products, the import of panel products has grown, as has the variety of their source of supply.

The imports of medium density fibreboard (MDF) have been growing very rapidly. Similarly, imports of particle board, hardboard, insulation board and oriented strand board (OSB) are increasing rapidly. However, their levels are still very low and their continued growth will depend upon the desire of Japanese consumers to substitute away from plywood. American experience shows that, in house construction, these panel products can go a long way in substituting for plywood and, if the price level stimulates such a change, the market penetration of these panels could be significant (Barnes, 1998). Imports of these main board types in 1997 according to country of origin are shown in Table 4.3.

The imports of particle board over the past 5 years have grown by an average

Table 4.3. Sources of panel products

Source	MDF (tonnes)	Particle board (cubic metres)	Hardboard (tonnes)
Canada	8,838	220,552	
USA	5,443	40,315	
New Zealand	121,079	75,397	
Australia	17,005	11,247	3,408
Chile	29,289		13,005
Indonesia		135,617	
Malaysia	99,260	86,580	1,750
Thailand		23,227	7,572
Korea	13,635		21,079
Brazil			1,383
Austria		26,901	
Belgium		17,542	
Italy	23,044		5,014
Portugal	5,714		
Spain	4,997		
Other	11,201	35,439	1,083
Total	399,505	672,817	54,294

source: Araya (1998)

of almost 40% a year and MDF by 28.5%. Against this imports of hardwood plywood rose by only about 10% but coniferous plywood imports, while being much lower than hardwood ones, rose by 32.5% a year (Araya, 1998). With the exception of Indonesia, Japan is not a particularly significant importer of particle board or fibreboard for the countries mentioned in the table.

Pulp, paper and chip imports

Japan's paper industry mostly imports pulpwood and chips for the production of paper rather than importing finished products, less than 2% of the world trade in newsprint is accounted for by Japanese imports. But, on the other hand, FAO figures show the dominance of Japan in the international market for chips. Figures for 1994 indicate that Japan imported nearly 73% of total chip volumes. This has led to accusations of protectionism, inevitably from the USA. In response, as in the case of plywood, voluntary guidelines were set up in an attempt to boost imports. Also, companies are entering into joint venture agreements to initiate pulpwood plantations. This is an area which causes environmental concerns as natural forests and indigenous species are often replaced by fast-growing exotic species.

The sources of pulpwood and chips, in RWE in 1996, are shown in Fig. 4.5 which is taken from Japanese publications. The main supplying countries were

Fig. 4.5. Sources of pulpwood and chips

Australia, the USA and Chile and the relatively undiversified position of Japan is shown in that these three countries accounted for over 78% of all chip imports. On the other hand, the supplying countries were also relative undiversified. Japan accounted for 98.8% of the chip exports from Australia and 99.4% of those of Chile. The relative share of Australian exports has grown somewhat in recent years at the expense of domestic inputs. The other countries have more-or-less maintained their market shares.

Japanese paper makers have been diversifying their sources of pulp and chips as a result of the growing environmental concerns in their more traditional sources of pulp. Countries such as Brazil, South Africa, Vietnam, Argentina and even Thailand have been used as sources.

Concluding Remarks

This chapter has looked briefly at the overall import position of Japan. One figure that has not been commented upon, although it was mentioned in Chapter one, is the fact that Japan only accounts for around 13% of the total international trade in forest products. This figure is a lot lower than the numbers that have been mentioned above and indicates that Japan's trade is concentrated in the lower value end of the market. This is something that has changed, although not in the very recent past. Figure 4.6 shows the value of imports into Japan, from

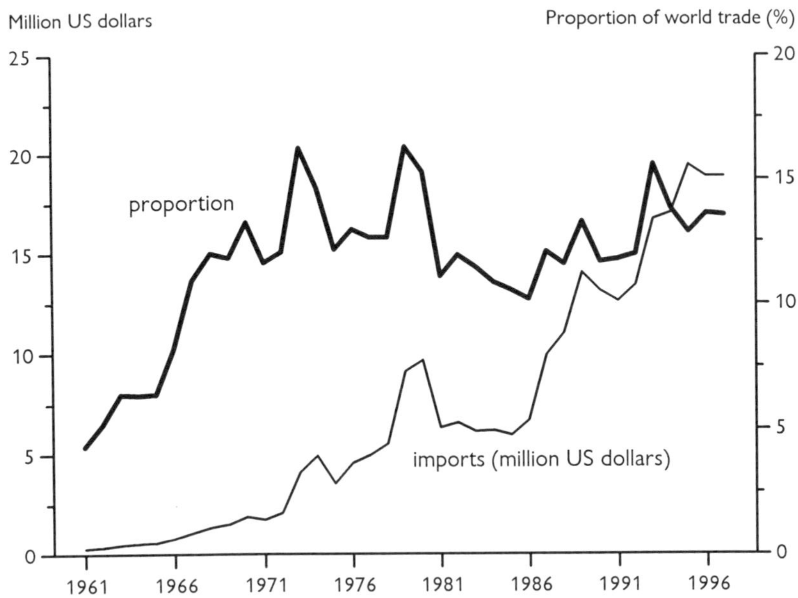

Fig. 4.6. The value of Japan's imports

FAO figures, and the proportion of total trade in forest products by value, again from FAO figures. The proportion of total world trade accounted for by Japan has remained fairly stable since the 1970s but the value of imports has risen rapidly, in line, of course, with total world trade in forest products.

The falling proportion of Japan's trade in areas such as log imports, which was mentioned earlier, implies that the increase in value of Japan's timber imports is at the expense of the lower value items and so Japan's trade is gradually moving towards the higher value end of the market. But Fig. 4.6 shows that the value of Japan's imports are increasing no faster than the trend in total world trade. Thus, there seems to be no specific trend for Japan to move to the imports of processed products any faster than other countries.

It is clear that a number of areas are of significance to the Japanese market. North America is dominant in the softwood trade, with New Zealand and Chile also increasing market share. Russia also has significant trading links with Japan and the breakdown within that country also shows that the importance of Japan is very great for the forest industries of the Far East. In the hardwood log trade Southeast Asia is the main area of importance with diversification of sources proceeding only slowly.

The next two chapters will look at these countries and attempt to assess their potential for maintaining or increasing the supply of timber and timber products to Japan.

Chapter five:

North American and Tropical Timber

Introduction

The discussion in the previous chapter attempted to put Japan's position in the international markets for forest products into some sort of perspective. Table 4.1 of that chapter showed that the USA held the largest market share and, in roundwood equivalent, accounted for more of Japan's forest products supply than did the Japanese themselves. Second in order of market share came Canada which, with the USA, supplied nearly 40% of all timber products consumed in Japan. The next most important grouping was the category *nanyouzai*, or tropical timber and timber products.

The figures showed that imports from North America were maintaining a more-or-less constant market share but that their composition was changing. Imports from Southeast Asia, the main source of tropical timber, were declining slightly but they still hold nearly 15% of the market. Given the importance of these two areas for Japan, it is probably best to begin the discussion of the foreign sources of forest products here. In this chapter the situation in North America will be looked at, followed by the position with regard to imports of tropical timber products.

North America

Japanese statistics often refer to *beizai* as a term which encompasses timber from both the USA and Canada. However, in this context these two countries cannot really be treated as being the same. Although the forests on their west coasts, which are the main areas supplying Japan, are contiguous and so similar in composition, the types of imports which have originated from these two countries differ. In general, imports from Canada have been in the form of sawnwood with only very limited log imports occurring in recent years. From the USA, however, logs and lumber have both been imported. Both countries also export panel products and chips to Japan.

The proportion of sawnwood imports from the USA has been increasing recently as a result of log export bans and reductions in harvests initiated because of ecological constraints. This was illustrated by the figures in Tables 4.1 and 4.2 in the previous chapter which showed that, despite the fact that Canada supplied over 13% of the forest products demand of Japan, it supplied less than one percent of the logs consumed in the country. On the other hand, the USA supplied over a quarter of the logs imported into Japan.

Imports from North America resumed very quickly after the Second World War and, as has already been noted, they were actively encouraged by the Japanese in order to stabilize timber prices. The result was the establishment of a sawmilling industry around the coast of Japan dealing mainly with logs from North America. The old-growth on the West Coast of the USA and Canada is particularly suited to the Japanese market as the timber has the strength characteristics which allow it to be used in construction and also has the aesthetic properties necessary for the more traditional segments of the housing market. Thus, pillars from North American timber have a long history of use in Japan and are in the market competing directly with those of sugi. The price differential between these pillars has now disappeared whereas, in the past, sugi was able to command a premium price.

As well as fitting in with the traditional markets in Japan, North America has been able to help shape them by its experience of wooden housing construction in its home market, especially in the area of two-by-four methods. As Chapter two showed, the two-by-four market is rapidly expanding in Japan and North American producers are well positioned to capitalize on this. Many Japanese companies are now importing *beizai* and marketing it as a product that is superior in strength and durability to the domestic species. The author remembers examining a show house in Tokyo where the salesman was adamant that the American timber used was much better than sugi and would last at least 50 years!

This, along with changes in preferences and cost considerations, means that the domestic market is more receptive to North American timber than to coniferous timber from other countries. More recently, the newer forms of structural panels, for example OSB, have started to be imported into Japan to supply the housing market and North America is well positioned to meet this demand.

However, the old-growth of America poses several environmental problems and this, along with the natural diminution of the growing stock stimulating the substitution of second rotation crops, means that the situation with regard to future export potential, especially of logs, is not easy to forecast. Also, once panel products rather than logs come to dominate the market, the American exporters expose themselves more directly to competition from other countries such as New Zealand and Chile. The logs from these countries have a relatively poor reputation in terms of strength and appearance, with wide annual rings and, if unpruned, regular whorls of knots making the timber weak. Thus, while their logs cannot compete in terms of quality, they would be able to compete if used for the production of panel products.

North American timber resources

In the Introduction it was stated that the Japanese often claim to have a 'culture of wood'. However, if anything, both the USA and Canada would have greater justification to laying claim to such a sobriquet. The per capita consumption of timber is greater in both countries than it is in Japan, with the figure for the USA being nearly double the Japanese one. However, because of the population difference, the demand for forest products is much greater in the USA than it is in Canada and so, whilst the USA is only 88% self-sufficient in forest products according to FAO figures, Canada produces more than four times the amount of forest products that it consumes domestically.

The United States' forest resources

The United States is a little over 30% forest covered and the West Coast area is somewhat more forested with forest cover reaching 40%. Forests are divided by ownership into private, federal and state lands. The distribution for the United States as a whole and the West Coast is shown in Table 5.1.

Table 5.1. The forest resources of the USA

			Private	
	Federal	State	Company	Other
USA total	39,116	14,100	28,513	116,394
Pacific coast northern	6,945	1,267	3,656	3,451
Pacific coast southern	3,512	180	1,327	1,820

All figures are in thousands of hectares.
source: USDA Forestry Service (quoted in *Mokuzai jouhou, 3/96)*

A feature of the West Coast timber resources is the prevalence of federal ownership of forest land. The overall figure for the Pacific coast region is nearly 30% whereas the national figure is only 13%. This is of significance because of the relative ease with which environmental pressures can be brought to bear on the timber harvests on federal land compared with those in private forests. Also, the relatively lower proportion of private forest means that any harvest restrictions on federal lands cannot be so effectively countered in the market place by compensating increases in the output from similarly located private forests.

However, in the USA, the difference between the physical resource and the economically and politically feasible harvest is greater than it is in many countries. After the end of the Second World War a strong demand for forest products, particularly in housing, created a large demand for timber and this stimulated an increase in output from national forests. At the same time harvests from private forests were beginning to drop and so the Forest Service 'shifted from its *custodial* management posture toward a more active policy of timber sales' (Adams and Haynes, 1989).

By the late 1960s the output from the forests of the west was, reportedly, reaching its sustained yield levels under the parameters of the management plans then in place. However, at the same time, these parameters were changing as environmental concerns increased. Large tracts were redesignated as wilderness and removed from the areas entering the allowable cut calculations and, in those areas not so designated, harvesting was modified to encompass non-timber outputs. As a result, since the late 1980s, timber harvests from federal forests in the Pacific Northwest have been falling.

One of the best-known manifestations of this increased environmental pressure which directly influences the available timber resource is the situation of the northern spotted owl in the old-growth of the West Coast. This is a case which will affect the projected prices of timber products and, therefore, the likely price of timber in Japan.

The Endangered Species Act was originally passed in 1973 as a part of public policy to protect biological diversity. As far as forestry is concerned, the main effect of this Act was to 'list' the northern spotted owl as 'threatened' in 1990 after a report by the US Department of the Interior Fish and Wildlife Authority concluded that the survival of the species would be imperilled by the continuation of the forest management practices then in use (O'Laughlin, 1992; Marcot and Thomas, 1997).

The designation of 'threatened' is the lesser of two possible categories, the more serious one being 'endangered', but the process of listing meant that a 'critical habitat' had to be defined. This is the area occupied by the species which contains the physical or biological features essential for its conservation. This listing also affects private forest owners as the Act prohibits the 'taking' of critical habitat anywhere in the USA. Unfortunately, from the point of view of the Japanese timber market at any rate, the habitat which was designated as critical for the northern spotted owl includes large parts of the old-growth stands of the West Coast and this, as will be seen, has affected the level of logging in the regions critical for Japan.

Naturally, this has had a 'knock-on' effect for the forest products industry. The sawmilling industry of the Pacific Northwest is around 50% dependent on logs from old-growth forests, although those working in the export industry are somewhat less so. Around a third of the industry in Washington and nearly two-thirds of Oregon producers are dependent on federal output (Weidenbeck and Araman, 1993). So the figures shown in Table 5.1 have effectively been reduced by such moves.

Canadian forest resources

Canada's forest cover is around 45%. Only 6% of the forests are in private hands with federal and provincial forests accounting for the remainder. Of the 417.6 million hectares classified as forest land about 37% is open forest land and around 56% is considered commercial forest. About 146 million hectares of this commercial forest is currently accessible by road, rail or water transport but 27 million hectares are excluded from harvesting by provincial policies protecting certain areas. Only an estimated 119 million hectares are currently managed for timber production. Forestry and the timber trade are of major importance to the economy, producing one in 16 jobs either directly or indirectly.

The main areas of Canada which supply the Japanese markets are again those in the west. The forest resources of the three westernmost provinces are shown in Table 5.2 and the predominance of provincial control is clear. Also, it is obvious that the species composition of the forests 'improves' to the west with more of the coniferous timber which finds favour in the Japanese markets. The forest industries are more important for these provinces than for Canada as a whole and in British Columbia the sector provides about 10% of the jobs.

Table 5.2. The forest resources of Canada

		Ownership (%)			Composition (%)		
	Forest area	Provincial	Federal	Private	Conifers	Broad-leaves	Mixed
Canada	417.6	71	23	6	67	15	18
British Columbia	60.6	95	1	4	89	3	8
Alberta	38.2	87	9	4	44	33	23
Saskatchewan	28.8	97	2	1	39	36	25

Area figures are in millions of hectares.
source: http://www.bcstats.gov.bc.ca

From the Japanese point of view, the published figures for the allowable cut and annual harvest are somewhat worrying. The annual allowable cut (AAC) for British Columbia is given as 68.1 million cubic metres while the harvest in 1995 was 74.5 million cubic metres. The situation for the other provinces in Table 5.2 is somewhat less critical. In Alberta the AAC is just over 23 million cubic metres against a harvest of just over 20 million and in Saskatchewan the AAC is 7.1 million cubic metres and the harvest only 4.2 million. However, in the last two provinces the greater proportion of the AAC was accounted for by the less desirable hardwood species (http://www.nrcan.gc.ca).

These figures need a little care in their interpretation. While output from national forests may exceed the AAC from year to year, on a 5-year basis the output of timber should equal or be less than the AAC. But, government figures for the AAC do not include private lands whereas the harvest figures include output from private forests. However, as the overall percentage of private forests is small and in British Columbia it is only 4%, the output and AAC figures suggest that no major increase in log output is really likely from these provinces. Indeed, as in other countries, the policy statements of the Canadian government now make frequent reference to the wider set of products that forests can provide other than timber. The result is that, like the USA, some areas of old-growth will be subject to use restrictions or withdrawn totally from forest management (http://www.bcstats.gov.bc.ca).

Timber output and Japanese imports

As has been seen, Japan is a significant market for North American timber. In 1995 the Japanese accounted for nearly two-thirds of all coniferous log exports and 47% of all coniferous lumber exports from the USA. It is far less important in the hardwood timber trade, taking only 10% of hardwood log exports and about the same proportion of hardwood lumber exports from the USA (*Mokuzai jouhou*, 7/96).

Japan is the second largest export market for forest products from Canada after the USA but its position is still significant, taking around 12% of the total exports of coniferous lumber. By far the most important region of Canada, from the Japanese point of view, is British Columbia. In 1996 just under 91% of the total forest product exports (by value) from Canada to Japan originated from British Columbia. The figures for lumber and softwood plywood were higher, 95.4% and 91.8% respectively.

Not only is British Columbia important for Japan but Japan is also of great importance for the province. In 1994 Japan accounted for 24% of British Columbia's forest products exports of Can$14.9 million, with the USA taking first place. The next province to the east, Alberta, shipped 20% of its forest product exports to Japan, but the overall level of exports was much lower, at Can$1.9 billion. Saskatchewan, the 'next' province, exported only 6% of its total of Can$399 million forest products exports to Japan.

The imports of logs and lumber from the USA and Canada taken together are shown in Fig. 5.1 and display the normal pattern. The data are taken from Japanese sources and the sawnwood series is in direct measure.

Most notable in the data illustrated is the steady increase of lumber imports and the decline in log imports which became most marked in the late 1980s and 1990s. The fact that most of the logs are imported from the USA means that the line labelled 'logs' in the diagram can be taken to refer, more-or-less, to United States' exports.

The log series shows a similar amount of variability as the total import figures but the lumber imports seem to avoid this short-term variability. The sudden recent drop in imports is also clear. This reflects the drop in log exports from the USA and a drop in lumber imports, mostly from the USA but also, to a lesser extent, from Canada.

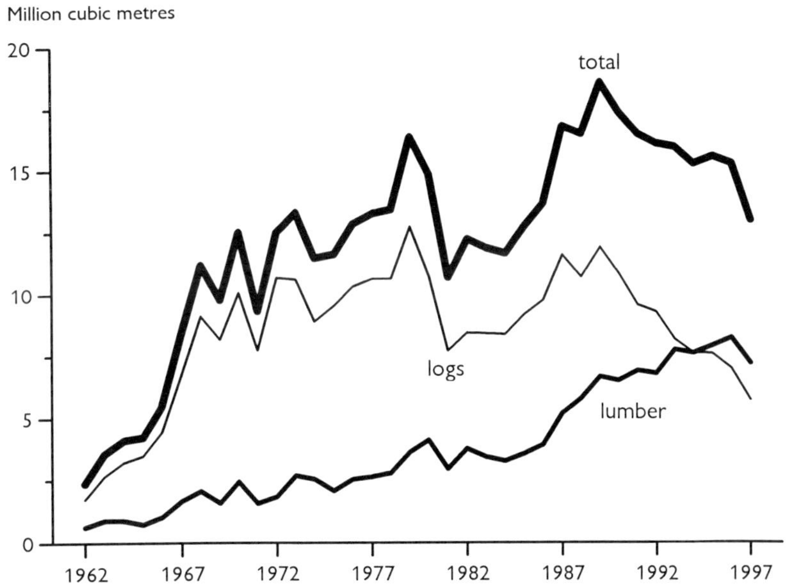

Fig. 5.1. *Beizai* log and lumber imports

The United States

The decline in log imports from the United States seems to be a purely supply-side effect. The total log exports of the United States declined over the final years covered by the Fig. 5.1 but during that period the proportion of coniferous log exports going to Japan increased somewhat. The Japanese seem to be responding to supply restrictions by competing for the remaining log supply and bidding the price up. A price premium for logs destined for Japan has long been paid but now figures for coniferous log exports from the States of Washington and Oregon indicate that this premium is showing a tendency to increase. As a result, an increasing proportion of the United States' declining log exports is shipped to Japan.

This is shown in Fig. 5.2 which presents the proportion of the Douglas fir log exports from Washington and Oregon that were shipped to Japan. This is the main species imported by Japan from these states and it accounts for between 70% and 80% of Japanese log imports. The proportion taken by Japan has increased despite the decline in the absolute level of exports during that time. A similar pattern is seen with hemlock exports where the proportion of exports from those States accounted for by Japan has risen from 82% to 95% over the same period. The main competing countries, Korea and China, have both seen their import volumes of these species decline and the proportion drop.

The price differential in Fig. 5.2 shows the unit price of Japanese imports as

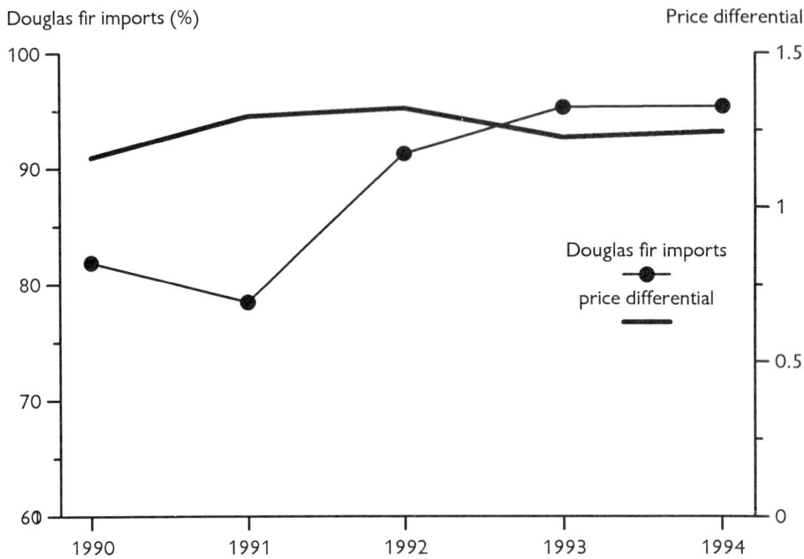

Fig. 5.2. Douglas fir imports from Washington and Oregon

a proportion of the unweighted average of the Chinese and Korean prices. Not only is this differential always greater than one but it rises in the early part of the 1990s. This suggests that the discussion in Chapter four concerning the potential for competition from Korea and China in the log market is somewhat less than the aggregate figures there suggested. These two countries are buying relatively lower price logs and Korea has, until recently at least, been relatively more active in the market for hemlock rather than that for Douglas fir (*Mokuzai jouhou*, 9/95).

The decline in log exports from the United States is a manifestation of the tendency 'to gradually and incrementally limit softwood log exports from western forests' (Sedjo *et al.*, 1992), which were first restricted in 1968. In that year annual log exports from federal lands in Washington and Oregon were limited to around 1.75 million cubic metres (350 million board feet using the conversion ratio of 5 cubic metres per thousand board feet from Perez-Garcia, 1995). Later the area to which the quota applied was extended to the whole of the West Coast. In 1974 the export of logs from federal lands in the west was completely prohibited. Therefore, all the exported logs since the 1974 restrictions have originated from private, state and local government land. In 1990 further legislation reduced softwood log exports from the west by a further 3 million cubic metres (600 million board feet) by allowing Washington State to export only a quarter of the harvest from state-owned land and banning all log exports from state-owned forests in Oregon. These bans effectively increased the price of those logs which were exported and served also to divert some exports to domestic processors.

As was shown in Chapter four, the position of Japan in the non-tropical hardwood timber trade is much less significant and this applies to imports from the USA. Also, the timber industry of the Pacific Northwest apparently sees hardwood species as being inferior to coniferous ones and output consistently runs at a much lower level than coniferous log production (Briggs and Buhler, 1989). Exports to Japan cover sawnwood for furniture manufacture and chips for pulping. Recent data suggest that the export of hardwood-based plywood to Japan is beginning, although it is still at a much lower level than exports of coniferous plywood. However, unlike the market for coniferous lumber, where North American exports can expect a price premium in the Japanese market, the hardwood market is far more 'normal' and so competition from Chile, China and possibly the hardwood resources of the Far East of Russia is more likely to drive prices down.

The hardwood timber trade is an area that has been subjected to less modelling effort than the coniferous trade, and Chapters seven and eight here are no exception to this rule. However, a model developed by Hansen and Luppold (1992) suggests, uncontroversially, that the demand for non-tropical hardwoods in Japan depends on price and GDP.

Canada

The reduction in log output from the United States and the diversion of logs to the internal market should have meant the American industry could have taken advantage of the market in Japan for softwood lumber. However, it seems that the American lumber manufacturers have not been able to do so and the main beneficiaries have been the Canadian producers. Figure 5.3, taken from Japanese data, shows the volume of lumber imports into Japan from Canada and the United States. The pattern is not the usual one of more-or-less stable demand but rather an increasing demand is shown. However, the increase in total volume imported is almost entirely as a result of increases of exports from Canada.

Like the USA, environmental pressures to reduce logging are being felt in Canada. However, the roundwood harvest has been increasing on a national basis over the years, although the harvest from British Columbia seems to have more-or-less stabilized. The figures for Canada and British Columbia from 1970 to the present are shown in Fig. 5.4. The proportion of the total harvest accounted for by British Columbia held fairly steady but, then, since the 1990s it has started to fall. More importantly for Japan, British Columbia's harvesting activities have been moving from the coastal areas to the interior. The proportion of the total roundwood harvest in the interior was only 45% in 1970 but this has continuously risen and now stands at nearly two-thirds.

The effect of this is to increase the costs of the timber harvested. Delivered wood costs in British Columbia have been rising by about 20% a year, although

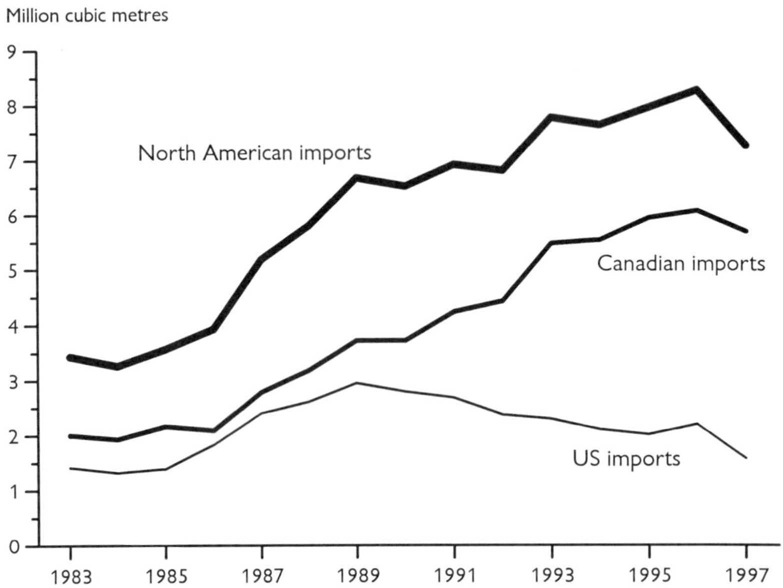

Fig. 5.3. Lumber imports from North America

most of this has been as a result of increased royalties and stumpage costs. However, if these costs are ignored, the logging costs in the coastal region are rising at about twice the rate of those in the interior as are roading expenses (http://www.for.gov.bc.ca).

Another factor that this shift heralds is the reduction in quality of the growing stock. Whilst the increasing costs might mean that Japanese importers would be willing to move further east for their imports, the forest composition becomes less and less favourable, from a Japanese point of view. As Table 5.2 showed, whilst British Columbia's forests are predominately coniferous, those of the other provinces are more-or-less half hardwoods.

Future prospects for imports from North America

The United States

For a long time, the potential output of the Pacific Coast has been the subject of analysis and 'for nearly three decades forestry experts have been projecting a decline in Pacific Northwest timber harvest levels in the 1990s' (Sample and Le Master, 1992). In 1974 the Forest and Rangeland Renewable Resources Planning Act of 1974 (RPA) called 'for periodic long-range assessments of timber supply and demand to provide information for stewardship and management decisions in

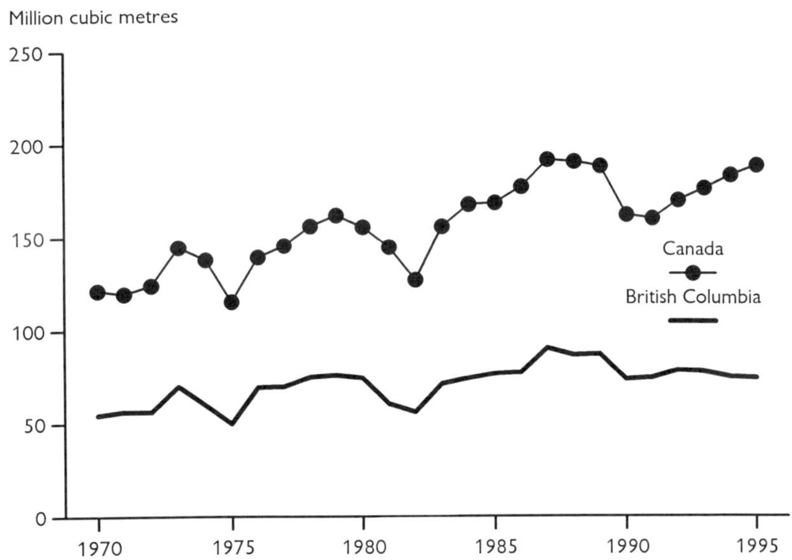

Fig. 5.4. Canada's and British Columbia's roundwood harvests

both the public and private sectors', a situation similar to the requirement on the Japanese Government to produce estimates of the demand and supply of its timber products (Haynes and Adams, 1992).

The last of these analyses was carried out in 1989, before the reductions in output caused by the listing of the northern spotted owl could be included (Haynes, 1990). The base scenario for the public harvest contained in the forecasts assumed a 2-3% growth rate in GNP and a continuation of the forest management policies of the 1970s and 1980s.

The projected levels of output for this scenario are shown in Fig. 5.5 which combines FAO output figures with the projections reported in Haynes and Adams (1992). The increase in the projected harvest levels of both hardwoods and softwoods is clear, as is the fact that the softwood harvest in the 1990s has not met forecast levels and that of hardwoods has exceeded it. The underlying distribution of output concealed in the aggregate figures in Fig. 5.5 were not encouraging from the Japanese standpoint. It was expected that the harvest from the Pacific Coast areas would fall by about 5% over the forecasting period with the increase in projected supply coming mostly from the Southern regions. From the Japanese point of view this would represent a detrimental change in the species composition of the harvest, with the old-growth Douglas fir and hemlock being replaced by timber from second rotation pine forests. Log exports from Washington, Oregon and California were projected to decline from their record levels in the late 1980s by nearly a third by the turn on the century and then

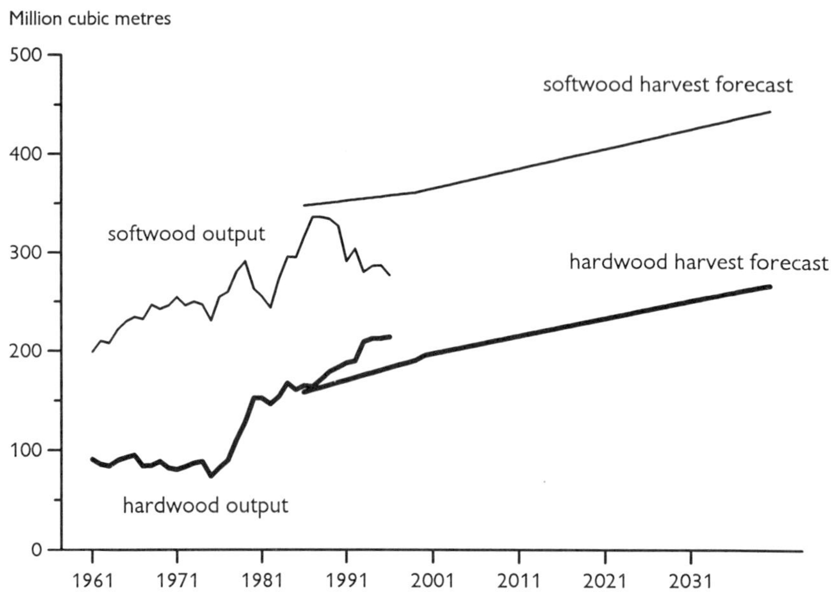

Fig. 5.5. Projected harvest levels from US forests

stabilize at that level. This would mean softwood log exports levelling off at around 12-13 million cubic metres a year. The implication, if Fig. 5.2 was foreseen, would have been a less than proportional drop in Japanese log imports as the premium in price increased.

Softwood lumber production was also expected to shift in line with the focus of the timber harvest, moving to the South and away from the Pacific Coast regions. In the forecasting period, softwood plywood and OSB output was expected to expand in all regions except the Rocky Mountains.

The prices under the base scenario of the 1989 RPA Timber Assessment are shown in Fig. 5.6. The Pacific Coast softwood stumpage price rises rapidly until 2020 and then begins to level off. Overall, the projected growth rate is 2% for the Pacific Coast region and 1.5% in the South. The more rapid early growth in stumpage prices results from the shortage of mature timber in the private forests of the Pacific Coast and South. In the latter stages of the forecasting period, timber stocks increase and harvests can grow in line with the maturing timber. The assumptions made in the forecasts are that efficiency improvements will go a long way to ameliorating the growing scarcity and lower quality of the timber harvest. This is shown by the fact that the prices of softwood lumber and plywood are expected to grow at a far slower pace, only 0.7% and 0.5% respectively until the year 2040, although most of the growth comes in the earlier years before the increase in output from the South can make its presence felt.

Of course, one of the factors which has upset the forecasts in the previous

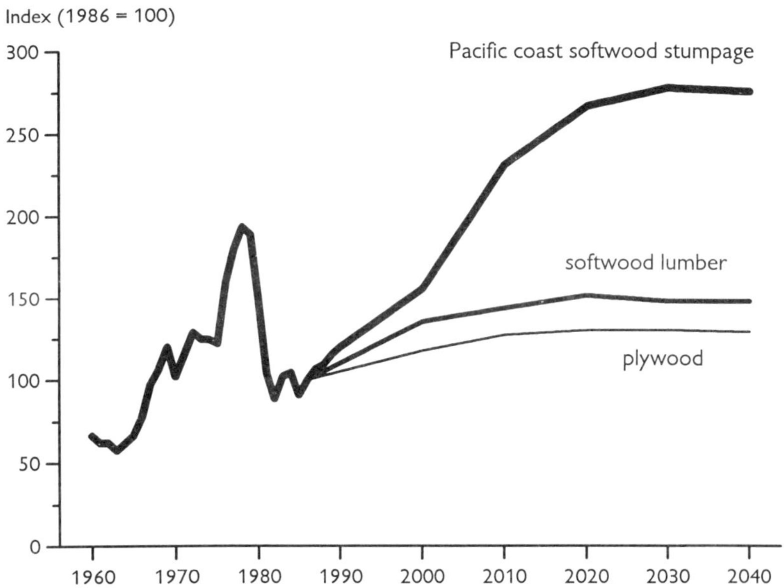

Fig. 5.6. Projected prices under the 1989 RPA Timber Assessment

two figures is the reduction in output from federal lands as a result of the conservation measures outlined in the previous section.

However, scenarios other than the base one were investigated and one such was a 'lower public harvest' scenario which assumed a reduction of around 40% in harvesting on public and private lands in the west. Large increases in stumpage prices resulted with corresponding increases in the softwood lumber price. Overall, softwood lumber prices grew annually by around 1% until the year 2040 compared with the base scenario figure of around 0.7%. In the period until 2010, however, the lumber price grew by 2% instead of around 1.5%.

Also, other changes are affecting private forest owners in the important Pacific coast areas. For example, Schallau and Goetzl (1992) report on measures to constrain management options on private land which have recently been instituted by a number of states. They report on a re-assessment of the RPA forecasts conducted by the American Forest Resource Alliance which simulated the effects of spotted owl legislation as well as other factors such as the preservation of more old-growth areas, the elimination of below-cost timber sales on federal lands, the protection of the habitat of more species and the banning of harvesting on roadless areas in national forests. As a result, they report that output in the Pacific Northwest would be 26% lower than the RPA forecast in the year 2000 and around 15% lower in the following decade.

Significantly, these forecasts included reductions in the output from the south of the USA instead of the compensating increases in output reported in the RPA exercise. Also, the forecasts included a reduction in lumber output in the USA as a whole, compared with the RPA forecasts of rough stability. The lumber prices in their model increase by 200% in the early years of the next century compared with increases of only around 50% for softwood lumber in the RPA base scenario (Haynes and Adams, 1992). In many ways this represents the 'worst case' from the view point of timber output, not really surprising given the provenance of the forecast, as the authors are both in the forest products industry.

The effects of the reduction in harvests in the west coast areas are expected to continue. Brooks (1995) states that, 'in the Japanese markets, reductions in shipments of North American timber can be expected to result in higher prices and adjustments in the sources of supply.' The argument that Brooks presents suggests that overall the level of shipments from the Pacific Northwest will fall, with log shipments falling more rapidly than lumber ones. However, for Japan the suggestion appears to be that there will be a substitution into lumber at the expense of logs and that the imports of lumber will continue the increasing trend shown in Fig. 5.1. Also, the sources of log and lumber imports will move to Russia, New Zealand and Chile (Sedjo *et al.*, 1994).

A lot of comment in the USA concerned this possible international relocation of timber harvesting as a result of the environmentally driven decreases in supply in the USA itself. The thrust of most of the arguments is that by reducing the supply from North America the harvest will, to a certain extent, shift to other parts of the world such as tropical forests or Siberian areas, where the environmental consequences of harvesting might be considerably worse than those of logging in the USA (Sedjo, 1995). However, whilst it is likely to be relatively easy to shift the source of supply in most countries, in Japan the resistance of the consumer to changes in species composition means that the process of changing the source and, therefore, the species of timber is going to be somewhat slower. This is already apparent in the prices paid by the Japanese and the increase in their market share for Douglas fir and hemlock from Washington and Oregon which was shown in Fig. 5.2.

New Zealand and Chile are ideally placed to exploit any decrease in North American timber harvests. However, as the quotation of Tanaka (1997a) in the following chapter shows, the perception that old-growth American logs and New Zealand radiata pine are direct substitutes is not one that would find much support in Japan. The outcome for the competition from other sources depends on the increase in panel products and, therefore, the rising share of two-by-four and *bunjou* housing which was discussed in Chapter two.

However, the proposed reductions in timber harvests in the Pacific Northwest mean the softwood logs harvested in the area might be diverted from export markets to the USA. Thus, mill owners in North America, given the likely cost advantages, might be able to outbid Japanese mills for the decreasing regional log supply. But Brooks (1995) concludes that such effects will be small and limited to coastal mills. However, he argues that lumber imports into the USA might become more important and, significantly for Japan, states that by the mid-1990s lumber shipments from Alaska had been diverted from the Japanese market to the rest of the USA.

In a study of this effect using the Global Trade Model developed at the Center for International Trade in Forest Products (CINTRAFOR), the effect on world trade of reductions in the Pacific Northwest and Canadian harvests was investigated. Two scenarios were examined. In the base case the timber sales from the Pacific Northwest were those proposed prior to 1990. In the so-called 'preservation' scenario a reduction in sales of 13 million cubic metres from public lands in the Western United States and 20 million in British Columbia was included (Perez-Garcia, 1995). The overall effect is that there is a supply shift for United States and Canadian timber and this induces a price increase. The price of sawlogs in the Pacific Northwest and in Japan are shown in Fig. 5.7.

The effects of these price increases are twofold. First, overall demand will fall and, secondly, production in other areas will increase partly to compensate for the drop in supply from the USA and Canada. In the overall results, softwood production falls by 33.3 million cubic metres, demand is reduced by 37.5% of the drop in production from North America so that the remaining 62.5% has to be met by increases in output from other regions. Whilst a lot of the compensating output comes from the USA, especially the southern areas, about one-fifth of it is supplied by increased European production and 2.1% comes from an increase in Japan. This is an extra output of 700,000 cubic metres a year or a 3% increase in

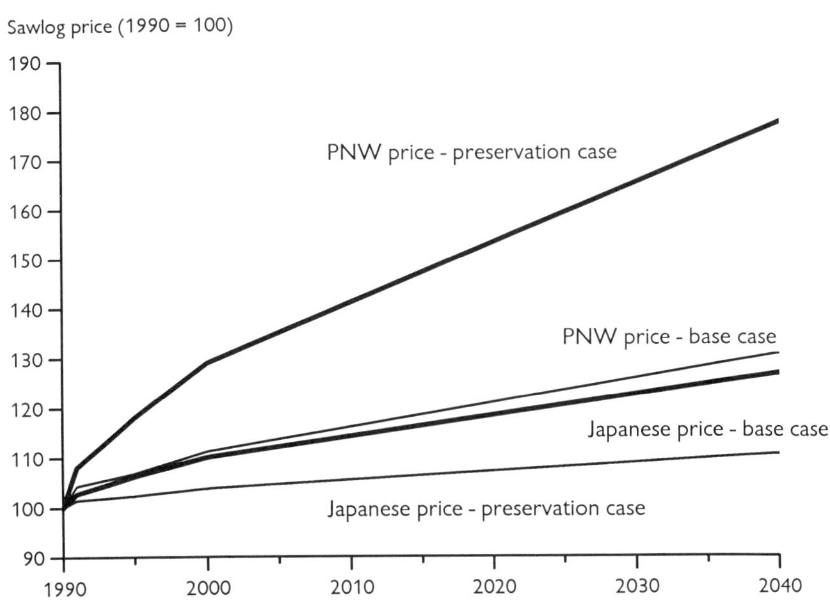

Fig. 5.7. Price projections of the CINRAFOR model

Japanese log output. This figure is not very much removed from the increase in output that would be implied by the model that the Japanese produced and is outlined in Chapter seven. The rise in price would increase the economic group ratio, which was discussed in Chapter three, by about 2 or 2.5% compared with the figure that would occur under the base scenario.

The results also show that the price of timber in Korea increases by somewhat more under both scenarios than it does in Japan, although starting from a lower base. So the implication seems to be that the competitive position of Korea vis-à-vis Japan increases somewhat. Indeed, some research already suggests that sawnwood manufacturers who target the Pacific Rim in general for their exporting efforts do better than those who look to just Japan (Gottko and McMahon, 1989).

A recent set of forecasts from FAO for the Pacific Rim timber trade paints an altogether more optimistic view of the United States' supply potential (Zhang *et al.*, 1997). These forecasts, based on a spatial equilibrium, market-clearing model for trade in the region, suggests that output from the USA will remain more-or-less stable over the forecasting period, i.e. until the year 2010. American output increases by about 6% overall but the projected level of industrial roundwood exports to Japan increases from its present level of between 8 million and 9 million cubic metres to 16.3 million cubic metres. This growth of almost 100% is against a backdrop of increases in total industrial roundwood exports from the USA of only about one-third. Korea is able to increase its imports from the USA but not in the same proportion. China, according to these forecasts, loses ground. Although the text does not discuss the sources in detail, if the reductions in output from the Pacific Northwest are accepted, the implication is presumably that Japan is able to substitute pine crops from the southern States for West Coast timber with ease.

The price pattern against which this increase in American output is set, is a doubling by the year 2010. The price of sawnwood is not projected to increase so rapidly, with softwood lumber prices increasing by about 70% over the period or around 4% a year. This is more than that in the RPA forecasts shown in Fig. 5.6 and also more than that in the CINTRAFOR forecasts shown in Fig. 5.7. However, the resulting pattern of trade in the forecasts has the United States exporting only a little more sawnwood in 2020 than in the base year of 1992. In this case the Japanese are not able to increase their imports. The projected intake of United States lumber by Japan is only just over 2 million cubic metres compared with recent import levels which have been in the region of 2.2 to 2.3 million.

Commenting on the changes in output levels the authors of the FAO report state:

> 'Such growths could be materialized only if business continued as usual in North America. However, a recent .... study suggested that continued expansion of lumber exports from North America may be limited by

resource constraints and environmental issues in British Columbia and the U.S. Pacific Northwest.' (Zhang *et al.*, 1997)

The implication of these forecasts is that the proportion of lumber in Japanese imports from the USA would fall to only 11%, using unconverted units. This goes against the trend shown in Fig. 5.1 and the present figure stands at 22%. It would also require quite an investment in Japanese processing capacity if the imported logs are to be converted into sawnwood in Japan. It also seems to imply an increase, or at least stability, in the proportion of the housing market accounted for by traditional post-and-beam methods. However, Chapter two argued that the trend is towards cheaper methods of housing where panel products can more easily be used.

From a different perspective, projections by Alig and Wear (1992) suggest a 4% drop in the overall forest area by the year 2040, mainly as land is lost to urban development. Some land is forecast to return to forestry as a result of excess capacity in the agricultural sector. Most of the loss is projected to be by small forest owners and so the effect on output is likely to be less than the overall area figure predicts. However, there is a change in the location of forests which the research suggests. With land in the West Coast area being redesignated as wilderness, Alig and Wear argue that the main plantation investment opportunities are in the South. This, of course, would represent the same species change that the RPA assessment implied.

Already, the Japanese seem to be anticipating this change in the centre of gravity of American logging. Articles explaining the situation in the southern States have been appearing in trade journals, although they usually emphasize the role of the production of panel products in the area (*Mokuzai jouhou*, 3/96; 5/96). Comparisons with the timber of New Zealand and Chile are often made and so it is doubtful if the timber from such areas could substitute perfectly for the West Coast timber in post-and-beam style housing. On the other hand, the increasing use of panels in Japanese housing means that the production shift to this area which many of the forecasts imply could still allow the USA to maintain a dominant presence in the Japanese timber market.

Canada

In Canada, as Table 5.2 showed, the private forest area is relatively small and the timber harvest is mainly determined by the policy of the provinces in allocating forests to industry and in determining the annual allowable cut (AAC). Each province regulates and controls the harvest rate on provincial Crown lands through the AAC, which determines the volume of timber that forest companies are permitted to harvest annually from a particular area over a specified time. The method of calculating the AAC varies from province to province and the calculations are based on the extent of the forest land area, the growth rate of trees, losses due to fire and pathogens, accessibility and economic and environmental considerations.

The AAC changes through time as each of the factors above varies and in this way, the calculated AACs go some way to reflecting the desires of society for environmental preservation. At the moment around 60% of the total AAC for industrial roundwood is in hardwoods whilst most of the demand is for coniferous species. This is less of a problem in British Columbia with its predominantly coniferous growing stock than it is for most provinces, but there is public pressure for greater weight to be put on the environmental outputs of forests, especially the old-growth natural forests which would find a good market in Japan. The trend of the coniferous allowable cut for all of Canada is shown in Fig. 5.8 (http://www.nrcan.gc.ca).

The AAC for British Colombia represents a little more than a third of the Canadian total and in the recent past the annual harvest from that province has been more-or-less equal to the AAC. Some companies in the province are importing timber from Alberta, the Yukon and Alaska to make up for local shortages.

There is some flexibility in the harvest as a result of harvests from private lands. The harvest from these forests has averaged about 7 million cubic metres but has occasionally reached 12 million when prices have been high. However, it is suggested that these harvests are not sustainable in the long run and so the Canadian Forest Service argue that:

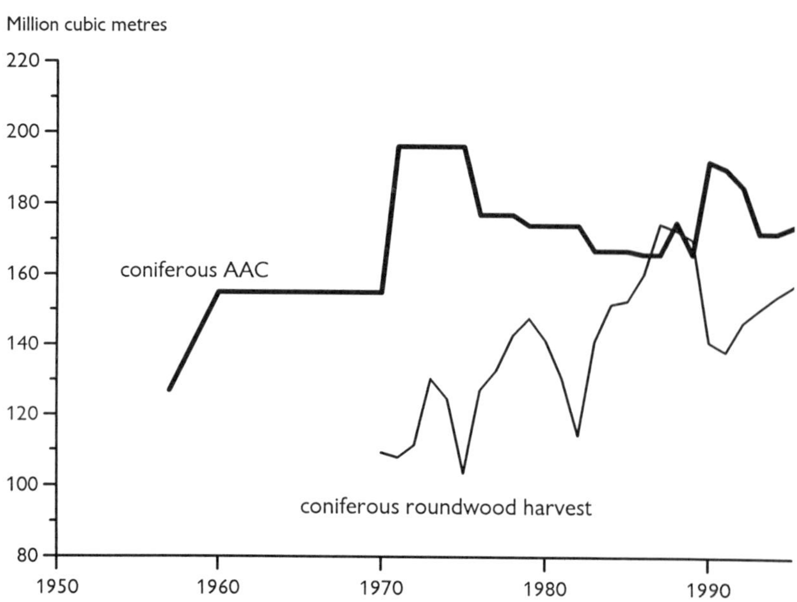

Fig. 5.8. The coniferous AAC and harvest in Canada

> 'Based on analyses to date, there could be a reduction in the AAC (softwood and hardwood) for regulated forest lands in BC from the current 71.3 million cubic metres to 60 million cubic metres by 2050, with a further decline to 57 million cubic metres in the long term.' (*The State of Canada's Forests*, http://www.for.gov.bc.ca)

Canada, maybe because of its log export ban and greater emphasis on processed products, seems to be doing better than the USA in the area of panel products. Canadian exports of particle board is increasingly gaining market share, and over this decade has increased from about a fifth to one-third of Japan's imports of the product. The imports from the United States have not shown such a rise and, although the absolute level is rising, the share of imports has dropped from around 16% to just under 6% (Araya, 1998).

The FAO report for regional trade discussed earlier in the section on the USA (Zhang *et al.*, 1997) suggests a somewhat steadier increase in the output of coniferous sawnwood from Canada but a quadrupling of log exports. The latter allows Japan to increase its present imports of logs from just over 100,000 cubic metres to about 2.2 million in the year 2010. The sawnwood imports show a more modest rise. The report gives figures for the increase as being from 4 million cubic metres in 1992 to 4.8 million in 2010. However, as Fig. 5.3 shows, this level of imports has already been surpassed. Again, like the forecasts for the USA, these predictions favour log imports and production at the expense of sawnwood.

Concluding remarks

The overall conclusion of most of the forecasts is that log output is likely to fall and prices rise partly because of the environmental factors mentioned but also because, for the first time, North America is facing increasing demand without a reserve of high quality coniferous timber to fall back on. These trends in price and output will be reflected by similar, if somewhat muted trends, in the Japanese market. The FAO forecasts argue the opposite but would require the reversal of many of the trends in the Japanese timber market if the extra logs, which are projected to be imported, can be processed and find an end-use.

Some suggest that there will be a substitution away from conifers, the most affected forests being coniferous, into the use of hardwoods (Weidenbeck and Araman, 1993). Indeed over recent years the non-coniferous harvest in the USA has been increasing against a background of a falling coniferous one. This growth has not really been reflected in its log export composition, whereas it has been in its sawnwood export trade. However, it is unlikely that this substitution could be made into the area of traditional Japanese housing, and its market penetration will depend on its use in panel products. Also, from the producers' point of view, the 'window of opportunity' for hardwood lumber might be quite short before southern forests in the USA come on-stream with a renewed softwood supply

(UN-ECE/FAO, 1996).

Despite increasing costs in recent years, both the USA and Canada enjoy significant cost advantages when compared with Japan and it seems unlikely that this position will be reversed. The topography and possibilities to increase labour productivity in logging mean that the basic raw material will be cheaper in North America (Kato, 1981). But also, the relative size of the sawmills in the various countries favours North America. Japanese statistics tend to quote size of lumber mills in terms of the power consumption rather than volume throughput. However, Sakai (1992) makes a comparison in terms of volumes of logs handled and gives average figures of around 50,000 cubic metres for Washington and Oregon, 90,000 in British Columbia compared with only 4500 cubic metres for Japanese mills handling imported logs. The overall figure for all mills in Japan is only 1800 cubic metres, and as was seen in Chapter one, the average is forced down by the small size of mills dealing with domestic logs. So the cost advantage for lumber is twofold. The raw material is cheaper to procure and the processing is more efficient. This would seem to mitigate against the large increases in log imports which the FAO study forecasts.

Not only is this cost advantage felt in sawnwood, it will also translate into panel products. The cheaper raw material, and the high wages in Japan, mean that the production of panels will be cheaper in North America than in Japan. Indeed, the increasing technological improvements and changes to panel products is seen as a way to increase output at a time of decreasing harvest. In the United States logging residues are now less than 10% at the logging site in coniferous stands and the amount of roundwood needed to produce 1 cubic metre of lumber and plywood has fallen from around 2.3 cubic metres in 1970 to just under 2 cubic metres (UN-ECE/FAO, 1996).

So, whilst the environmental factors seem to mitigate against exports to Japan, the economic factors suggest that North America can maintain its position as a leading exporter to Japan.

Tropical Timber

The dependence of Japan on tropical timber has been declining but remains significant. The main trends have been a move to the import of processed products away from logs and a change in the sources of supply. Most of these movements have been induced by the supplying countries and price changes.

Tropical timber resources

Much has been written about the nature of tropical timber resources and the problems of deforestation. Only a cursory summary of the argument will be rehearsed here. Over the past decades, Japan has been the main importer of timber from tropical countries, what the Japanese refer to as *nanyouzai*. At the same time the area of forest in those countries has been diminishing and the areas which are still defined as forests are often degraded. Many argue that it is clear that these two events are causally linked.

The loss of natural, tropical forests is estimated by FAO to have been 14.63 million hectares a year over the period 1981 to 1990, a slight lower figure was originally estimated. The total loss of forest was less, at 12.26 million hectares, indicating an increase in the area of plantation forests. The loss of natural tropical forest was just under 1% of the total area of such forests. In percentage terms, the greatest level of deforestation occurred in the Southeast Asian countries which are the main sources of supply for Japan. The latest figures for the period from 1990 to 1995 show that both the absolute level of deforestation and the percentage rate are declining for the world as a whole. In Asia and the developing countries of Oceania the rate of loss of natural forests is now running at 1.14% a year with the highest figure being registered in mainland and insular Southeast Asia (FAO, 1997). The figures for individual countries of significance to the Japanese market are shown in Table 5.3.

Table 5.3. The state of Southeast Asian forests

			1990-1995		1980-1995
	Area in 1990	Area in 1995	Area	%	%
Burma	29,088	27,151	387	1.4	
Cambodia	10,649	9,830	164	1.6	
Indonesia	115,213	109,791	1,084	1.0	1.0
Laos	13,177	12,435	148	1.2	
Malaysia	17,472	15,471	406	2.4	1.8
Papua New Guinea	37,605	36,939	133	0.4	0.3
Philippines	8,018	6,766	262	3.5	2.9
Thailand	13,277	11,630	329	2.6	2.9
Vietnam	9,793	9,117	135	1.4	

Area figures in thousands of hectares.
source: *State of the world's forests*, FAO (1997)

The production figures for industrial roundwood for the important Southeast Asian countries, taken from FAO sources, are shown in Fig. 5.9. The diagram highlights the differences between the various countries in Southeast Asia. Both the Philippines and Thailand have seen falling timber production whereas Indonesia and Malaysia have witnessed increasing levels of log output. The question which the graph poses is whether the management of forests in Indonesia and Malaysia is *better* than that in the other countries in the sense that a levelling off rather than a downturn can be predicted for the future or whether the upward trends seen now are merely the harbingers of a future decline in output of the type seen in the Philippines or Thailand.

The rates of deforestation for Indonesia in the two periods covered by the FAO data in Table 5.3 show that there has been no let up in the reduction of forest area, although as the percentage deforestation rate seems to be unaltered

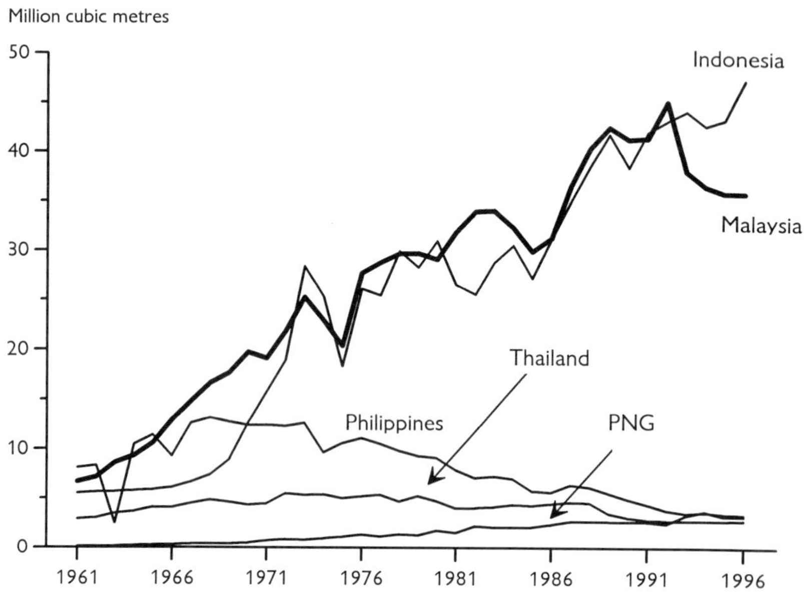

Fig. 5.9. Industrial roundwood production in Southeast Asia

the absolute loss of forest area is decreasing slightly. The figures for Malaysia show a drop in the rate of deforestation but, here, the location of the loss of forest cover is of significance. The forest industry of peninsular Malaysia differs from the islands of Sabah and Sarawak. Log exports and most of the deforestation will be found on the islands rather than the peninsular. The recent drop in log production in Malaysia shown in Fig. 5.9 reflects the lower harvests levels in both Sabah and Sarawak, with the Sarawak's harvests from its permanent forest estate now at the annual level of 9 million cubic metres. This is the level recommended by an ITTO Mission to Sarawak in 1990.

The two countries which are increasing their timber harvesting, Malysia and Indonesia, are also increasing their production of timber products. Figure 5.10, which is again taken from FAO data, shows the rapid increase in plywood production over recent years and the equally rapid increase in exports. However, despite this apparent internal log processing capacity, ITTO figures show that, compared with the Asian region in general, Malaysia processes a relatively lower proportion of its log harvest. Overall, Asian countries processed just over 90% of their timber harvest domestically in 1992 and this figure rose to 94.3% in 1996. The situation in Malaysia, whilst showing a rise, reached only 81% in 1996, up from 75.5% 4 years earlier. There has been a great deal of discussion as to whether the domestic production of plywood in Indonesia and Malaysia represents an economically efficient use of resources. Economic analyses suggest that greater profit could be achieved by the export of logs and that the conversion

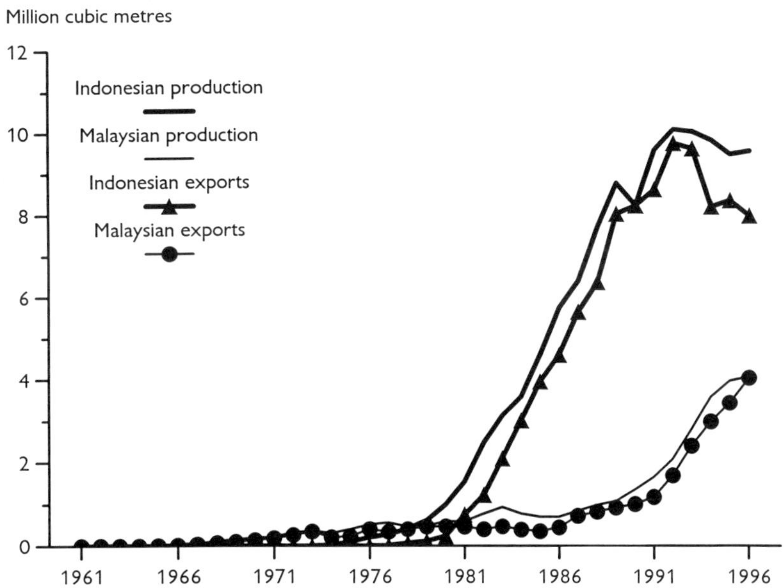

Fig. 5.10. Indonesian and Malaysian plywood production and exports

rate of logs is greater in Japan than in Southeast Asian. The conversion ratios of the plywood mills are often lower and, because the raw material is relatively cheap, there is less of an incentive to use it efficiently. The use of waste and lower quality logs is less than it might be if the raw material were processed elsewhere.

However, this might change with the increase in the production of reconstituted panel products. Again, there is some disagreement over the advisability of such production in Southeast Asia. The increased use of raw materials might mean more intensive, and therefore more destructive logging, over a possibly decreased area. The high-grading of natural forests, assuming no encroachment from agriculture, at least gives the forest a theoretical chance at recovery. In Sarawak there are plans to build a MDF plant in co-operation with the Japanese and plans were announced in 1997 for a joint venture with China to produce such a plant in Peninsular Malaysia (ITTO, 1997). Thus, economic and environmental arguments notwithstanding, the diversion of logs from the world market to domestic industry represents a 'fact of life' for Japanese importers.

Japanese tropical timber imports

Japan dominates the global tropical log market, importing more than three times the volume of the second most important nation in the market, China. Most of these logs are used in the plywood industry where the fine finish is valued, if only for giving concrete a smooth finish when the plywood is used as shuttering or *kon-pane*. Recently, as was shown in Fig. 2.9 in Chapter two, Japan has been increasing its imports of plywood as countries impose log export bans and the plywood industry of Southeast Asia develops. Japan is also the largest importer of tropical timber plywood.

Figure 2.9 also showed that Japanese domestic plywood production fell below plywood imports in 1995 for the first time. Japanese plywood manufacturers are increasing the proportion of softwoods used in plywood production and there are some non-tropical hardwood plywood imports now originating from North America. Also, techniques to allow the re-use of *kon-pane* are being investigated, although at the moment it is only used about two or three times before being burnt. This is of major significance as, at present, about a third of tropical timber in Japan is used for *kon-pane* (Sedjo *et al.*, 1994). Thus, an increase in its usage from three to six times would cause tropical timber demand to drop by about one-sixth. Several plywood manufactures are establishing joint ventures for plywood production in countries in the region. These factors, together with the declining supply of logs, mean that Japanese production of tropical plywood will continue to decline. Similar trends are noticeable in competing countries with Korean production of tropical plywood falling by 43% since 1993.

The source of logs for the Japanese market has been changing through time as Japan slowly diversifies its sources of supply. However, by far the largest proportion of logs still comes from Southeast Asia. Figure 5.11 shows the sources of Japanese tropical log imports, the figures being taken from various Forestry White Papers. The overall picture is one of decline but the only source of supply showing an increase is 'others'.

The main source of supply in the early years, the Philippines, supplied over 90% of Japan's tropical timber log imports in the 1950s but gradually Malaysia and Indonesia took over the role of main supplier. This change is reflected in the large deforestation rates that were registered by the Philippines and shown in Table 5.3. The Philippines rapidly became depleted of suitable timber and in 1973 a log export ban was announced which was to be phased in over the following 3 years. This was never implemented but in 1975 a selective export ban was imposed. In 1986 log exports were finally banned although exports from plantations are still permitted (Dauvergne, 1997). The movement away from the Philippines as a main source of supply was effectively made before these restrictions were enforced as the main Japanese trading houses looked elsewhere for their supply.

Indonesia played a major role in supplying Japan for a number of years. However, Indonesia has followed an explicit policy of encouraging the domestic production of plywood and other timber products and so instituted a log export ban in 1985. In 1971 Indonesia became the main source of tropical logs for the Japanese market which imported about three-fifths of the total log production in Indonesia. The large trading housing and their affiliated companies dominated

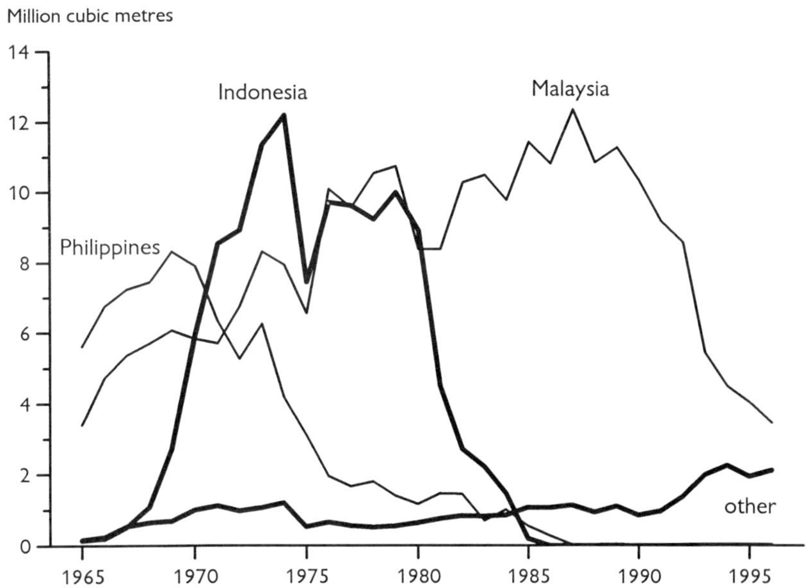

Fig. 5.11. Tropical timber log imports

this trade. In the early 1980s the Indonesian government began the implementation of the log export ban and Japanese imports began to decline, to be replaced by imports of plywood. In May, 1992, under the threat of action by the United States government, the export ban was replaced by export taxes which were of such a scale that, effectively, log exports ceased (Barbier *et al*., 1994).

The imports from Malaysia have been more resistant to events but even these began to decline in the 1990s, by which time they constituted nearly all of the Japanese imports. Peninsular Malaysia has had a log export ban in place since 1971 and so the imports into Japan come from Sabah and Sarawak. In 1993 Sabah imposed a log export ban which, apparently, came as a surprise to the Japanese plywood manufacturers causing log prices to double. Exports are now permitted on a limited basis. There are no plans to expand the forest industries of Sabah (ITTO, 1997). The output of logs from Sarawak is also falling but Japan accounts for the lion's share of its exports. In 1995 Japan imported 4,060 thousand cubic metres of logs from Sarawak which was down slightly on the previous year. However, this reduced figure still accounted for more than half of the State's log exports (*Mokuzai jouhou*, 96/7).

The increasing quantities of logs aggregated under the umbrella 'other' in Fig. 5.11 includes timber from Papua New Guinea and Burma in Southeast Asia and interest is being shown in Cambodia and Vietnam and other countries for log, lumber and plywood imports. These countries are developing a logging industry and their deforestation rates are increasing as the figures in Table 5.3

suggest. However, Papua New Guinea is, under current policy, hoping to reduce log exports by 10% a year until the turn of the century and then ban them altogether. Also, the IMF suspended loan payments to Cambodia in 1996, largely as a result of illegal logging and log smuggling. As a result, the country has a ban on log and rough sawnwood exports and it is looking to increase its production of all timber products including particle board and MDF.

Japan is diversifying further afield with significant imports now coming from Gabon, Cameroon, Ghana, Congo, Côte d'Ivoire and Zaire. These countries accounted for just over 2 million cubic metres of tropical logs in 1995 and nearly 2.5 million the year before. However, as a means of securing a long-term source of logs such a strategy could well fail. In 1994, the government of Cameroon introduced a law by which concessionaires can only export 30% of their annual cut as logs. After 5 years, this figure is to fall to zero, so that the entire volume harvested will be processed in Cameroon. In September 1995, Côte d'Ivoire announced a ban on the export of most logs in an attempt to boost its local wood processing industry and to protect its remaining forests. In Gabon, a gradual reduction of log exports is planned in the long-term as, at the moment, only a very small proportion of the logs produced in the country is domestically processed. A local processing rate of 50% of the harvested timber is envisioned for the year 2000 and 90% by the year 2025 (ITTO, 1997). So the policies in use in the Asian countries, of industrialization by encouraging their forest industries, are being repeated in other countries.

If the import figures in Fig. 5.11 are compared with figures for industrial roundwood exports in total for the various countries, a picture emerges of Japan taking a roughly constant proportion of all exports until a log export ban is introduced. For the Philippines the figure was around 80%, for Malaysia it still hovers around 50-60% and for Indonesia it was around 70%.

Figure 5.12 presents the end-use data for *nanyouzai* over the years. Most of this timber is used for plywood although a significant proportion of it is also used for lumber. Although the Japanese statistics do not break down this end-use by country of origin, in 1993 just under 80% of *nanyouzai* was used for plywood manufacture and about 7% for lumber. In the past an even greater proportion of tropical timber was used for lumber, in the past the figure has been as high as 30-40%.

The declining import of tropical logs has been matched by an increasing import of tropical timber-based plywood but not by an increasing import of tropical lumber. Here the use of tropical logs for the production of *kon-pane*, which was noted earlier in this chapter and in Chapter two, is of significance. Not only is *kon-pane* the largest end-use of tropical timber in Japan, it is also regarded as the most wasteful. The Japanese claim that the finish produced by tropical timber faced plywood is superior to that of other types of panel but, given the fact that very little visible concrete results from construction work and that it is usual to clad the concrete panels it seems a largely spurious characteristic of the concrete forming material. The reason for its use is probably

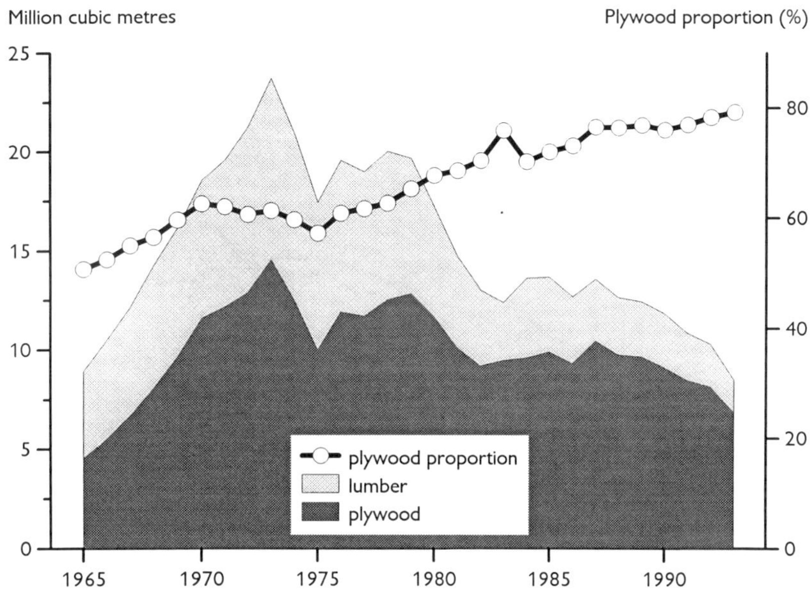

Fig. 5.12. The end-use of tropical timber imports

just cost. Most Japanese are now willing to state that the reason for the use of tropical timber in this area is its cheapness and that an increase in its price would see the substitution of other types of shuttering materials and other raw materials used in plywood manufacture. Indeed, in most other countries other panel products are used for concrete shuttering.

Part of the reason for the drop in the use of tropical sawn timber is the relative prices of the logs. Figure 5.13 shows the relative prices of sugi and lauan log prices. The relative price of lauan has risen from around three-quarters of the price of sugi in 1975 to its present level of about two-and-a-half times the price (comparing sugi *chuu maruta*, 14 to 22 cm diameter, 3.65 to 4 metres with unit prices for lauan saw logs, 60 cm minimum diameter, length over 4 metres). Similarly, as a use of tropical timber, plywood is becoming relatively cheaper in the sense that saw logs are becoming more expensive when compared with plywood logs. The dotted line in Fig. 5.13 shows the relative price of lauan saw and veneer logs. The former is becoming even more expensive compared with the latter.

Ball (1997) argues that the average thickness of plywood in Japan is increasing and this suggests that the production of *kon-pane* is also increasing as this has a thickness of 12 mm, greater than the average. He found the average thickness by dividing the FAO figures for volume of plywood in cubic metres by the reported Japanese figures in square metres. Whilst this is a very dangerous way to treat the statistics, there has been a statistically significant increase in the

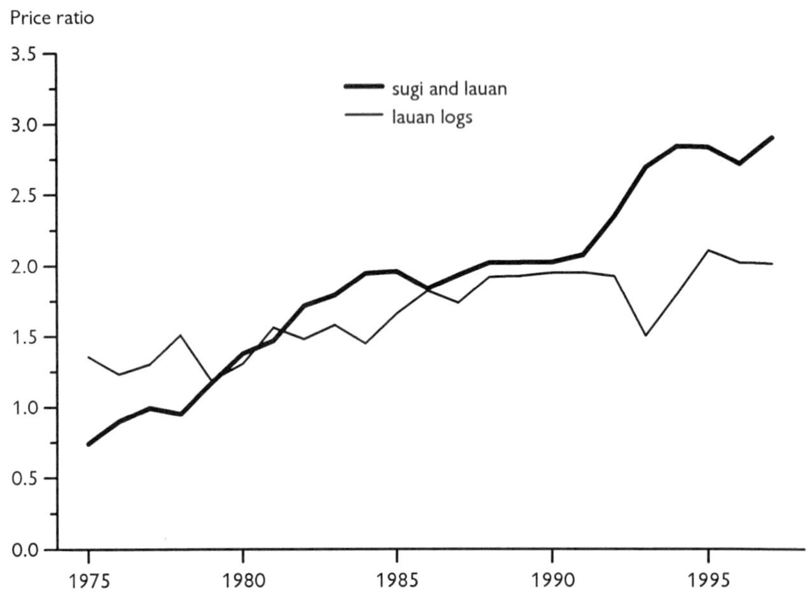

Fig. 5.13. Relative prices of tropical and domestic logs

thickness of domestically produced plywood over the period 1961 to the present but, interestingly, a slight decrease in the thickness of imported plywood.

Future prospects for imports of tropical timber

The area of forecasting the potential supply and demand for tropical timber is far less developed than it is for other areas, mainly because of the lack of reliable data and the fact that almost all of the data series 'move in the same direction'. Deforestation and GDP, production of timber products, agricultural output, population, are all closely related and so the ability to tease out the relative importance of the various factors is very difficult.

One attempt to model deforestation and trade potential is that of Grainger (1990). In his model deforestation is dependent upon population, per capita food consumption and yield per hectare. Under a high and low scenario, deforestation for the Asia-Pacific region was estimated to be between 0.4 million and 1.2 million hectares a year over the period 2000 to 2020. Fairly steady deforestation rates are predicted for Indonesia and Malaysia with rates dropping to zero for Thailand and the Philippines as these countries effectively reach the base levels of forest area that were assumed in the model.

Additionally, and importantly for this study, Grainger also simulated trends in regional exports of tropical hardwoods. Under his optimistic scenario the export of tropical hardwoods from the Asia-Pacific region has already reached its peak and is now declining, while exports from Africa and Latin America are set to peak in the early years of the coming century and decline rapidly thereafter.

This model is not dissimilar to that of Burgess (reported in Barbier *et al.*, 1994) where population and agriculture crop yield were also seen as influencing the decline in forest area. Significantly, Burgess also finds deforestation linked to per capita roundwood production, a 1% increase in roundwood production being associated with a 0.02% decrease in forest area. Unfortunately, Burgess does not seem to consider explicitly timber exports but, for the discussion above, it would seem that a drop in tropical timber output will translate more-or-less directly into a corresponding drop in exports, either of logs or of domestically produced sawnwood or panel products.

One possibility is the substitution of tropical hardwoods by other species, either plantation timber from tropical countries or timber from other sources. The motivating factor here would mainly be that of price and so the market for tropical timber would have to be working moderately efficiently for the increasing scarcity to be translated into increased prices. Given the risk attached to holding tropical timber stocks, either in the sense of a change in governments causing a shift of patronage and so a loss of access to timber rights, or the encroachment of forests by landless peasants or other forms of illegal logging, the ability to withdraw tropical timber stocks from production in anticipation of higher prices is that much reduced. This implies that the ability of the market to signal impending shortages through current price increases is limited and so quantity signals might be more efficient than price ones in this context.

This seems to be the position of many of the governments in question as moves are made to increase the output of processed products even though the profitability of such enterprises is not at great as the export of logs might be.

The FAO report on regional trade patterns (Zhang *et al.*, 1997) has log output from the Southeast Asian countries increasing slowly or staying approximately stable. However, Japan is able to capture an increasing share of this production with imports from Malaysia forecast to rise. In the original report the directions of trade statistics show an increase in imports from 8 million cubic metres in 1992 to 12.2 million in 2010. However, Fig. 5.11 shows that, rather than increasing, log imports from Malaysia have dropped and the latest import figures are just under 3 million cubic metres (Forestry White Paper, 1997).

The forecasts for plywood suggest a slight increase in output from Southeast Asia but almost entirely as a result of increases in Indonesian production which rises by about 21% over the study period from 1992 to 2010. Exports show a slower growth rate as, presumably, production is diverted to domestic markets. The main feature of their direction of trade figures for plywood exports is that China is able to compete successfully with Japan and increase its imports from Indonesia by more than the Japanese. Korean imports remain roughly constant over the period.

There are relatively few studies on the effect of prices on substitution of timber. Many of these are summarized in Barbier *et al.* (1994). With regard to sawn timber, the work of Cardellichio *et al.* (1989) is interesting in that it shows an elasticity of demand in Japan for non-coniferous sawnwood of 2.4 and for coniferous sawn timber of 0.7. This suggests that increases in prices of tropical hardwood sawn timber would lead to much greater reductions in consumption than would be the case for coniferous lumber. Also, it is possible to imagine that the reduction would lead to the substitution by coniferous sawn timber for hardwood lumber so that the two may be viewed as fairly close substitutes in end-use.

Concluding remarks

The International Timber Trade Organization (ITTO) argue that 'few of ITTO's Asian members have the potential to substantially increase log production sustainably' (ITTO, 1997). Also the avowed policy aims of the countries in the region is to increase the domestic processing of their logs. The same applies to some of the other regions of the world to which Japan has begun to turn for log imports. So the long-term forecast must be that log imports will decrease and imports of plywood and other panels will increase and that, increasingly, coniferous products will be substituted. Again, this will tend to relate to the housing market. The FAO regional forecasts (Zhang *et al.*, 1997) seem to argue the opposite, as it did in the case of North American production and trade, with output of logs from Southeast Asia not falling and exports in a number of cases increasing markedly.

One aspect of the potential to export is the level of absorption of the products by the domestic market and, therefore, the growth rates of the domestic economies of the Southeast Asian countries. The 1997 ITTO Annual Review states with regard to Malaysia:

> 'Domestic building activity is expected to continue to expand significantly for the next few years. ... The buoyant domestic construction activities are expected to increase domestic timber consumption, although current economic uncertainty may slow this somewhat.'

Given recent events in the region, the final sentence seems an understatement but the downturn in the economies could well put pressure on the governments to expand timber outputs as a means of increasing foreign exchange earnings at the short-term expense of domestic development plans. The extent to which domestic processing capacity is dependent on external capital sources will limit the moves to panel production. The crisis in many of the economies of the region suggests that inward investment is likely to be much reduced, in the short term at least. As yet the effects are not really visible in the data and so the overall result of the crisis in Southeast Asian markets remains a matter of speculation.

Conclusions

The two regions examined in this chapter are the main ones from which Japan derives its timber products. In the past, both regions have been rich in high quality, large-diameter timber with the exact characteristics that fitted the Japanese market, not least of which was a favourable price. As a result, the old-growth coniferous timber from the USA was able to find an outlet in the traditional markets in Japanese housing, and tropical plywood was sufficiently cheap and of such good quality that it was able to more-or-less exclude other panel products from the market.

All that has now changed and the price and availability of both sources of raw material are under threat. For many countries, where the visible characteristics of timber are of less importance, the shift to second growth conifers and panel products will be fairly easy. For Japan, however, the change will be more difficult and will depend on the increasing share of western-style housing making the timber market more 'normal' in nature. If Japanese preferences demand traditional timber this might be sufficient to drive up the price of sugi and hinoki and so cut back on the demand for timber products from these countries. If, on the other hand, such preferences are not as strong as some might think, and Chapter two suggested that this might be the case, the continued comparative advantage of North America and Southeast Asia means that their market share will be maintained, but that the composition of the exports will alter.

Chapter six:

Russian, New Zealand and Other Timber

Introduction

The sources of timber looked at in the previous chapter are those which have traditionally supplied the bulk of timber and timber products to Japan. In this chapter attention will be turned to what might be termed the minor sources of supply. Of these Russia, New Zealand and Chile are the main areas to be considered. These are similar in that they have been supplying timber which, in the main, is considered to be of somewhat lower quality than that originating from North America and the tropical countries. The futures faced by these countries are also very different. The market reforms in Russia have caused timber output to fall rapidly and have increased the power of environmentalists who might succeed in reducing future timber harvests once the market system there becomes more firmly established. On the other hand, New Zealand and Chile are both home to large man-made forest areas which are coming to maturity and both have a strong export orientation. As well as these sources, Japan is now beginning to turn further afield. Timber imports for Northern Europe have begun and these will be looked at briefly.

Russia

Russia has long supplied Japan with timber and the species composition of its forests in the Russian Far East is similar to that in Hokkaido. The disputed islands to the north of Hokkaido were Japanese before the Second World War and, at that time, were an important source of timber for Japan. Imports from the Soviet Union in the inter-war years reached a peak of between 2 and 3% of the Japanese market but tailed off in the 1930s and stopped altogether during the Second World War.

Although trade with the Soviet Union re-opened in 1954, imports resumed later than those from North America with the first significant imports beginning in the late 1950s. Imports were influenced by a typhoon which blew down 25 million cubic metres of timber in Hokkaido in 1954. It took 3 years to market the salvaged timber and this partly accounts for the low level of timber imports from Russia in those early years of resumed trade. But because the volume of timber blown down saturated the Hokkaido market it had to be shipped to other areas of Japan. So the typhoon effectively established a market for ezomatsu and todomatsu (spruce and fir) on Honshu. As these were the species that the Russians wished to export the foundations were laid for the larger-scale import of timber from the Soviet Union (Kakizawa, 1994).

Imports from Russia, or *hokuyouzai* as they appear in Japanese texts, are mostly coniferous logs and Russia holds a market share of around one-third of this market, as was shown in Fig. 4.2 in Chapter four. It is far less important in sawnwood and other timber products. In round wood equivalents, Russia now holds 5% of the total Japanese market (Forestry White Paper, 1997).

Russian forest resources

Land classified as forest occupies 886 million hectares of the Russian Federation, giving Russia a forest cover of around 45%. Within this category of land, the actual forested area is estimated to be 763 million hectares (Nilsson, 1997b). The total growing stock of these forests is over 80 billion cubic metres of which 79%, according to UN/ECE figures, is coniferous (UN/ECE, 1997). This figure is somewhat higher than previous Soviet ones and the International Institute for Applied Systems Analysis (IIASA) places the proportion of coniferous forests at 72% (Nilsson, 1997a). This forest base gives Russia an extraordinary timber increment and, therefore, theoretical annual allowable cut (AAC). Estimates of this vary but figures of around 800 million cubic metres a year are often quoted. This roughly corresponds to a rotation length of 100 years and a stocking at maturity of 100 cubic metres per hectare.

The Russian Far East (see Fig. 6.1) is, not surprisingly, the main area supplying the Japanese market. In 1995 77% of total Russian log imports came from this region. From the Russian point of view, the Japanese market takes the lion's share of exports with nearly 80% of all exports from the Far East going to Japan (Kondrashev, 1997a). As the Far East exports about 14% of its total log output, it is clear that the Japanese market is very significant and it would be useful to look at the situation in the Russian Far East.

Soviet figures showed that the total forest resource of the Far East covered 238 million hectares with a total growing stock of 20 billion cubic metres (Yakunin *et al.*, 1987). More recent figures stand at 270 million hectares and 21 billion cubic metres respectively (Kondrashev, 1997a; 1997b). In area terms, nearly two-thirds of these forests are classified as being mature or overmature, meaning that the age of the stands is over 100 years (for coniferous

Fig. 6.1. The Russian Far East

and mixed forests). The volume of timber in the mature and overmature age classes was estimated by the Soviets to be 13.5 billion cubic metres. This gives an average growing stock of only 93 cubic metres per hectare which compares to the figure for the whole of the former Soviet Union of 124 cubic metres (Vorob'yev *et al.*, 1979).

The lower figure for the Far East reflects the relatively poorer quality of the forests there. Figures in Vorob'yev *et al.* (1979) illustrate this point and they are summarized in Fig. 6.2. For the Soviet Union as a whole 10% of forests were on sites of 'high quality' and only 20% on 'low quality' sites. The comparable figures for the Far East were 3% and 29%. The Far East had the lowest proportion of high quality sites and the highest proportion of low quality sites of any region of the former Soviet Union. Also, excluding the former republics of Soviet Central Asia and Georgia, the Far East had the lowest proportion of fully stocked stands.

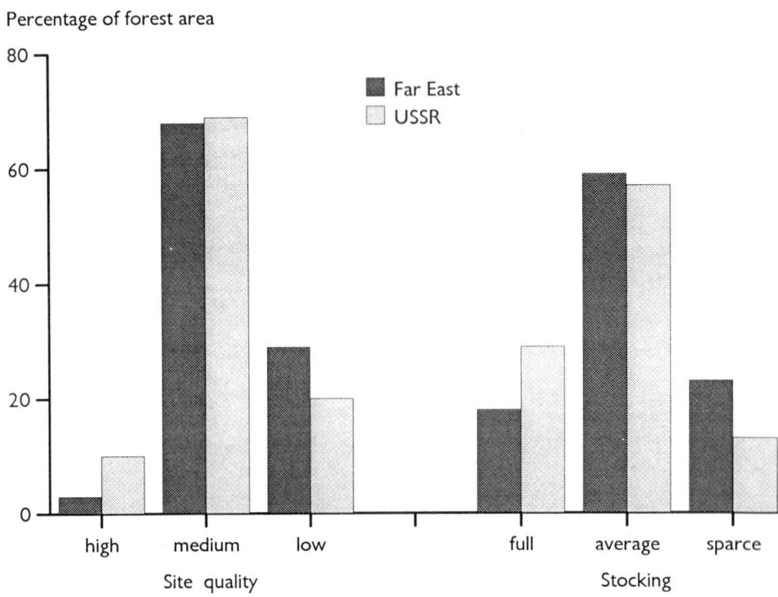

Fig. 6.2. The quality of the Far East's forests

Not only is the overall composition of the Far East's forests often worse than the average, the trend has been towards deterioration of the forest stock. Nilsson (1997b) presents figures which show that the growing stock of Russia's forests as a whole has increased by around 13% over the period 1961 to 1993. However, the figures for Siberia and the Far East from the 1980s to the present day show a decrease of about 3%, with the growing stock returning to the level of the late 1970s.

As well as these factors, the uses to which the forests are to put reduce the overall harvest potential. Even in the Soviet days, 10% of the forest area was earmarked for Group One, which is that group of forests to be managed for environmental protection purposes. Of the Group Three forests, which are those managed for timber production, only a little over a half were allocated for actual exploitation (Yakunin *et al.*, 1987). In recent years reforms have been aimed at reducing the environmental damage done by activities such as logging, but the Russians have been slow to put such reforms into practice. However, it is clear that increased environmental concerns are likely to limit timber output in the region (APRENet, 1994). In this sense the inaccessibility of the forests of the Far East is somewhat of a benefit. Only about 15% of Siberian roads have hard surfaces and are useable all the year (Nilsson, 1997b). This acts to preserve some areas but, conversely, to encourage overlogging in those areas which are accessible.

The main species of the Far Eastern forests is larch, making up about 55% of the forests which were designated for exploitation under the old Soviet system (Yakunin *et al.*, 1987). Kondrashev (1997b) puts the figure for the whole of the Far East at 62%. Larch is less popular in Japan than spruce or fir because of its resinous nature which means that it is difficult to work and unacceptable where the appearance of the timber is important. Around 70% of the mature and overmature larch stands are said to be affected by heart rot (Krankina and Ethington, 1995). As Chapter one showed, larch is used a lot in packaging and plywood manufacture. Also it is used in sills for wooden housing because of its durability. However, Cardellichio *et al.* (1990) argue that the price differential has already accommodated the preference for other species and its durability could even be a benefit in future market developments. They conclude that the large volume of larch does not pose an impediment to the future development of the forest resources of the Far East.

Overall, therefore, the quality of the forest, location and accessibility are factors which drastically reduce the possible harvest. Russian figures show that the AAC for the Far East region is around 100 million cubic metres (Kondrashev, 1997b) which is significantly greater than the present output of the region, which stands at around 10 million. (Kakizawa, 1998). The AAC of the main area in the Far East supplying Japan, Khabarovsk, stands at around 20 million cubic metres (APRENet, 1997) which would imply that the present level of imports by Japan are not only sustainable but that the timber is there to supply Japan's foreseeable timber import needs.

However, the overall figures for the allowable cut are really only relevant to the long term. Other factors are likely to be of more significance in the short run.

Timber output and Japanese imports

Figure 6.3 shows the pattern of Russian timber production since 1961. The data prior to 1991 are 94% of the output of industrial roundwood for the whole of the Soviet Union, as reported by FAO. This was roughly the proportion of the USSR's timber output for which the former Russian Republic accounted. The figures after 1991 refer to Russia. Of course, *perestroika*, the fall of communism and the beginnings of a free market system have all affected Russia's forest industries, causing significant falls in output and this is clear in Fig. 6.3. However, the graph also shows that output in the former Soviet Union peaked in the mid-1970s. After that output began to fall and, in part, this fall reflected the stagnation that gave rise to the reforms that eventually saw *perestroika* and the fall of communism.

Japanese imports from Russia peaked in the 1970s and then showed a slight decline, partly because the price of Russian timber rose to equal that of timber from North America and sawmills began to change to American timber. However, exports to Japan have not been greatly affected by the recent events in Russia. The dramatic drop in domestic production in Russia has been mirrored by a fall in domestic demand for timber and the collapse of some of the other export

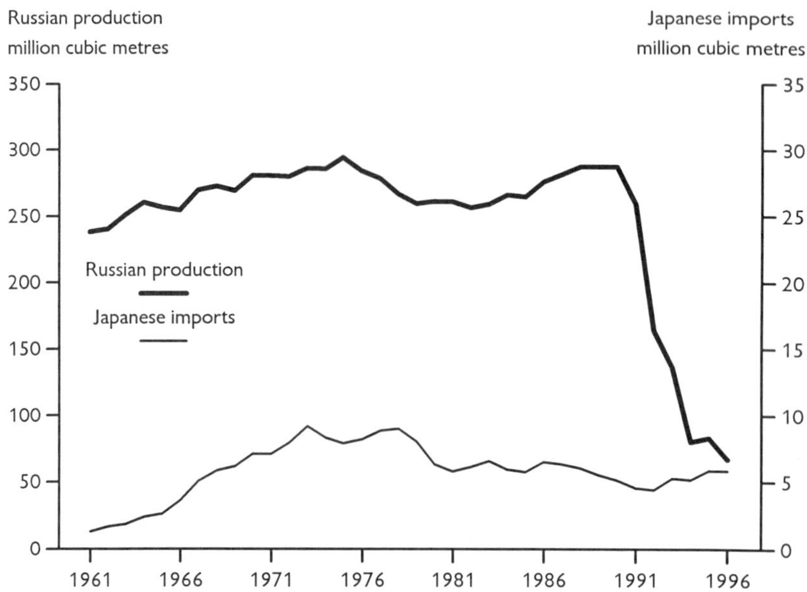

Fig. 6.3. Russian timber production and Japanese imports

markets such as the former Eastern Bloc countries. This has allowed export levels to the Pacific Rim area be maintained (Lochkarev, 1995b).

The robustness of exports to Japan is not a manifestation of a slower relative decline in the output of timber from the forests of the Far East. The decline in Russian output has been almost exactly mirrored by the fall in output from the Russian Far East (Kakizawa, 1998). So imports into Japan have been maintained despite the turmoil in the Russian forest sector. Indeed, it is likely that they have been maintained because of that turmoil. The Japanese market represents a source of hard currency earnings which is likely to be favoured over other markets for the timber of the Far East. So, in a similar way to the situation in North America shown in Fig. 5.2 in the previous chapter, Russia seems to be a favoured market for Russian timber products.

Table 6.1 shows the destination of the exports from the Far East as a proportion of the total exports from the region. Japan dominates the market for logs and lumber but is (to one decimal point at least) absent in the other market areas. The table also shows a lack of 'head-on' competition between Japan and the other Asian importing nations. Only Korea and China are significant in the log market but Korea tends to import more hardwood logs and lumber whilst Japan's imports are largely coniferous.

Table 6.2 shows the sources of timber for the Japanese market within the Far East region and it is clear that the Khabarovsk region is the most important one although recently the output from the Khabarovsk region has declined somewhat

Table 6.1. The importance of Japan for the Russian Far East

	Logs	Lumber	Plywood	Pulp	Newsprint
Japan	75.0	79.1			
Korea	14.7	0.5		46.6	
China	10.0	9.1	100	53.4	
USA	0.2				
Hong Kong	0.1				
Taiwan & Singapore	0.1	6.1			
Malaysia		2.8			
Uzbekistan		2.2			
Indonesia		0.2			
Vietnam					100

All figures are percentages.
source: *Mokuzai jouhou* (7/97)

quicker than average (Kakizawa and An, 1997). Also, Table 6.2 shows that most of the exports directed to Japan were in the form of logs. This is partly because of the relatively low quality of the timber products from Russia and partly because of the lack of capital investment in the Far East. The first imports of lumber were recorded in 1964 and constituted 1.4% of total imports from Russia. The trend in this proportion is shown in Fig. 6.4 which gives lumber imports as a proportion of lumber and log imports combined (both in absolute measure rather than RWE) and compares this series with the North American situation which was discussed in the previous chapter. The proportion only passed 3% in 1988 and has only recently begun to rise to around 7%. The contrast with the situation for North America is clear.

One favourable aspect of this feature of Japanese imports is that it suggests that the disruption caused by the reforms in Russia is less likely to be detrimental to Japan's position than to that of other countries which import more processed timber products. The decline in maintenance and investment which has been caused by the reforms has clearly affected logging but the output of processed

Table 6.2. Imports from the Russian Far East

	Logs	Lumber	Chips	Pulp
Primorskii	469	100	30	
Khabarovsk	2445			13.1
Amur	600			
Kamchatka	30	5		
Sakhalin	670	6		23.4
Yudaya	5			
Magadan		3		
Total	4219	114	30	36.5

Log and lumber figures in thousands of cubic metres, chips and pulp in thousands of tonnes.
source: Kondrashev (1997b)

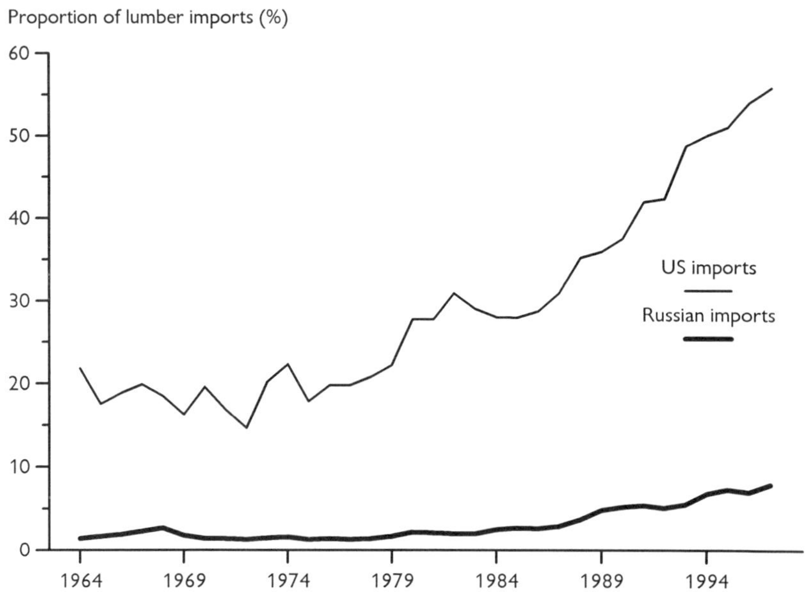

Fig. 6.4. Lumber imports as a proportion of total log and lumber imports

timber products will have been more adversely affected. Kakizawa (1998) presents figures which show that the falls in forest products in the Far East are even more dramatic than the drop in log production. Output of sawnwood has fallen to 16% of its 1985 level, paper to 6% and plywood to just under 3%.

Part of the reason for these drops is the way in which reform has been handled. The break-up of what previously were integrated concerns might work well in economies with established means of creating the necessary trading links. In Russia, however, the reliance on the external markets for flows of intermediate products is not going to be successful until that external market system is operating. The centralized planning system meant that links between producers and consumers were very indirect and so there was little to build on in the move to a market economy.

The low proportion of lumber in exports to Japan causes concern to some Russians. The reasons for the low level of lumber exports are given as technological (Levina, 1997). The Russian mills are, it is reported, simply incapable of producing the high quality lumber which the Japanese market requires. This will be an ever-growing problem even if the increase in two-by-four housing reported in Chapter two continues. How quickly this situation changes will depend largely on external capital.

However, while it might be tempting to predict that increased investment in the Russian Far East will lead to extra imports of sawnwood, there are problems with this prognosis. In an interview reported in Levina (1997), the director of

Sakhalinlesprom states that it is 'paradoxical' that it is less profitable to supply lumber than logs. In the Soviet days this would have been acceptable, even desirable, because of the need for foreign exchange. There still seems, however, to be a conflict between the desire to earn a profit, designated in part in roubles, and in earning foreign exchange in the gross sense, a point raised by Cardellichio *et al.* (1990). Whether profit or revenue generation will dominate is not clear.

From the Japanese point of view, the problems faced have hardly changed over the years. Barr and Braden (1988) reported problems with species assortments and log grades in the exports to Japan, with difficulties caused by the amount of larch in the shipments. Such problems remain with grading systems and species assortments still at the top of the agenda in bilateral meetings (Matsuo, 1995). Also the length of time required for logs to reach the Japanese market, especially in summertime, means that they are often discoloured on the way. One of the features of logging in the areas which supply Japan is its seasonal nature and the use of ice roads. Without all-year roads it is probable that such degrading will continue and, when the nature of the Japanese market is considered, access to markets with premium prices will be impossible.

The main species imported by Japan are ezomatsu, todomatsu, karamatsu, akamatsu and benimatsu (*Pinus koraiensis*). Over the past 10 years or so the proportion of these species has varied and this is shown in Table 6.3. The figures for 1994 show that the largest proportion of log imports is taken up by todomatsu and ezomatsu, together accounting for nearly half of all imports. Benimatsu has dropped in importance. The reason for this trend is that previous harvesting has reduced standing volumes and moves have been made to preserve the environment of tigers found in benimatsu forests (Cardellichio *et al.*, 1990). The proportion of karamatsu has remained constant at around 28% but this is significantly lower than its preponderance in Far Eastern forests.

Table 6.3. Russian saw log imports

	Total	Ezomatsu & todomatsu (spruce & fir)	Karamatsu (larch)	Akamatsu (pine)	Benimatsu	Hardwood
1986	5940	2550	1703	659	347	681
1990	4310	1720	1269	400	96	825
1994	4474	1746	1225	1092	37	374

All figures in thousands of cubic metres.
source: Matsuo (1995)

The end-use of Russian timber is usually reported in the aggregate and not broken down according to species. Almost all, however, is used for lumber and plywood, with only a little going for pulp. The proportions for 1996 stood at 74% for sawnwood, 22% for plywood and the rest for pulp. This is a drop on the 1965 proportion for sawnwood of 91.6%, mainly because of the increase in the use of *hokuyouzai* for plywood. The species mostly used for this is larch. However, the same problems emerge here as in the use of Russian logs for lumber. The lack of consistent sorting of logs by species and grade means that, whilst Russian larch is the main imported species used in plywood production, its use is limited to the Japan Sea coast. More distant mills would have the expense of disposing of, or selling and transporting, unwanted logs that did not meet the standards needed for plywood.

The breakdown of end-use of Russian timber is presented in more detail by Kakizawa (1994). He reports that, until the 1970s a lot of Russian timber was used in the production of squares for use, not as the visible pillars, but for structural components not usually seen. However, because of rising relative prices and the lack of sorting by grade and species, this end-use has largely been taken over by North American timber and lumber. Now, according to a survey undertaken in 1991 and reported by Kakizawa (1994), spruce and pine imports are mostly sawn into small sectional lumber and used for non-structural items. A number of reasons are given for this. One is the relative cheapness of the timber. Others are its softness and ease of working and the good quality of the raw material. Clearly, the first and last of these reasons are subject to change as the forest resource declines and costs in Russia rise. Larch, on the other hand, because of its high resin content and relative durability, is still cut into squares and used for sills.

As was noted in Chapter two, the housing market in Hokkaido is somewhat different from that in the rest of the country, and this is reflected, to a certain extent, in the use of Russian timber. Spruce, which is a white wood, has been used for visible columns in housing in that area. This is partly because it is a native species in Hokkaido and partly, so inhabitants of other parts of Japan would have one believe, because it is cheap and people in Hokkaido worry more about costs than their compatriots to the south.

Future prospects for imports from Russia

Figure 6.3 showed that Russian timber production began to decline as early as 1975. Two views of this fall were available in the English language literature and the reasons which were given are still of relevance to forecasting today. First, Blandon (1983) argued that labour productivity was not rising at a sufficient rate to offset the increasing scarcity of labour and declining resource in the forest-rich areas of Russia despite the equipment being introduced to mechanize forestry. Thus, the increases in output seen in the 1960s and early 1970s could not be expected to continue. Such a pattern of events did, indeed, seem to materialize. The proportion of total labour employed in repair and maintenance work continued to increase, exceeding 50% in 1988 and the percentage of the various machines used in logging being useable at any given time being no greater in 1988 than it was in 1960, and in some cases actually lower (Yagodnikov and Mikhailov, 1991). This is something which is a result of the severe conditions in Russian forestry and the poor reliability of Russian technology. The latter could be overcome by purchases of foreign equipment or improvements in Russian equipment and is an area which joint ventures between Japan and Russia can and do address.

Barr and Braden (1988) concentrated more on the forest resource and their argument was that poor planning, a creaming of the available forest resource and the harshness of the environment meant that the ability of the Russian forest to sustain its level of output was questionable. They suggested that much of the forest of the Soviet Union had been mined rather than exploited in a sustained yield framework leaving a landscape 'devoid of trees or with forests lacking commercially viable stands'. However, the more recent work of IIASA tends to contradict this view. Nilsson (1997a) argues that the increase in growing stock which has occurred makes 'it difficult to justify the premise that Russian forests are disappearing'. Chelyashev and Abe (1997) state that the area for forest in all parts of the Far East increased between 1966 and 1993.

Both of these analyses were incompatible with the figures which were used by IIASA in its Global Trade Model (Kallio *et al.*, 1987). In its base scenario it assumed that timber removals from the USSR would increase gradually from 360 million cubic metres in 1985 to 430 million in 1990 and would stay at that level until 2010, a forecast which events have shown to be optimistic.

Factors affecting the export potential of Russia's forests

Given the changes in Russia over the last decade, the forecasting of production and export levels is especially difficult. It is possible to list, however, the factors which are of importance and an attempt will be made to do so here.

The accessible growing stock depends on prices, investment in infrastructure and environmental considerations. The latter is becoming increasing important. There are ten natural reserves in the Primorskii area and there are plans to create more in the areas which serve Japan. Overall, however, the theoretical AAC is such that supplies to Japan, in the long term, are biologically feasible.

The reduction in output described above and summarized in Fig. 6.3 has had the 'beneficial' effect that a pool of labour is now available in the forest sector. Unemployment in the forestry sector is very high, at least 17% for the country as a whole and 42% in the Far East (Granåsen *et al.*, 1977). It seems that, in purely numerical terms at least, there is no labour shortage as was emphasized by Blandon (1983). However, if output is to be stimulated by increases in technology and especially the import of foreign technology, the problem of the skill level of the pool of unemployed workers becomes a question.

Most of the investment funds will, as is normal, have to come from internal cash flow and this is still a problem for many logging concerns. *Roslesprom* stated that 37% of the enterprises in the forestry sector were on the verge of bankruptcy, and most of these were probably in the logging and lumber sector (Rinaco, 1996). Inflation is seriously reducing investment. According to Lipman (1994), investment in new production capacity in timber production in the last half of the 1980s was only half the level in the period 1976 to 1980, and investment in sawnwood production was only a quarter. The author is more familiar with the state of the capital stock in European Russia but, there, a great deal of investment is required to bring it up to western standards and the impression gained by the writer supports Lipman's figures.

However, large-scale investment is largely dependent upon foreign capital. Backman (1996) reports that Russia is second only to Iraq in terms of country risk for foreign investors, a ranking which current events will have done little to alter. The illiquidity of the stock market and the difficulties of joint ventures all reduce the likelihood that there will be a substantial influx of capital in the near future. What activity there is in the Russian stock market seems, at the moment, to be limited to the pulp and paper sector rather than logging. Also, despite the privatization of many industries in Russia, the forests remain in government hands, although long-term leases are possible.

The attitude to exports is not really clear. It will be influenced by exchange rates, internal domestic growth and the need for foreign exchange. As the internal economy recovers so too will the internal demand for timber. Given the relative strength of nationalistic feelings in Russia today and the lack of a long history of free market economics with its emphasis on importing and exporting according to economic opportunities rather than ideas of self-sufficiency, the words of Backman (1996) that '[w]ith the present environment in Russia, it is unlikely that domestic consumption in Russia would be allowed to suffer through unrestricted export of wood fiber' ring true. However, Backman and Waggener (1994) suggest that the central authorities would sacrifice domestic consumption to maintain exports at roughly their current levels. The question of exports is likely to be more political in Russia than in other economies for some time to come.

If past figures represent a form of long-run equilibrium to which the Far East will return when the reforms have worked their way through, about a quarter of the timber output of the region will be exported. In the 1980s this was the proportion exported with only 15-20% going to other areas in Russia. The rest was consumed within the Far East. Now, however, the proportion exported has risen to around 40% and it is likely that this proportion will drop. Whether this means a decrease in exports or just a relative fall depends, of course, on the output of the region (Kondrashev, 1997b). However, a recent report on the Tumen River Area Development Programme for the UNDP argued that 'Siberia and the Russian Far East have a relatively small population, so the growth of domestic demand would only slightly affect the use of Siberian natural resources' (APRENet, 1997). To a certain extent, the lack of infrastructure is an advantage to Japan. The forests of the Far East are probably 'nearer' to Japan in a cost sense than they are to the markets of Siberian and European Russia.

Despite the fact that internal and external prices in Russia are often subject to market imperfections, it seems to be the case that the profitability of exporting logs exceeds that of lumber (Lochkarev, 1995a). This recalls the discussions on the desirability of exports of logs or timber products from developing countries.

However, figures from Jaako Pöyry (APRENet, 1997) suggest that the cost per cubic metre of producing sawnwood to Japanese specification in Russia is about half the cost of doing so in Japan using imported logs. Provided that wage inflation in Russia does not erode this substantial differential and investment in equipment, which allows the production of good quality lumber in Russia, proceeds, the trend to import lumber rather than logs, which Fig. 6.4 shows has begun, should continue.

Of course, one of the factors which will influence the Japanese ability to import will be the development of other markets for Russian timber in the area. China and Korea are both likely to grow over the long term and begin to compete with Japan for imports. As Table 6.1 showed these two countries both import significant logs quantities from the Far East, although they are more important in the pulp market.

Although Korea tends to import hardwood logs and lumber rather than softwoods, their import of softwood lumber is set to increase to between 1.5 and 2 million cubic metres in the next century as hardwood resources in Russia dry up. China's imports of softwood lumber are also expected to double to around 1.2 million cubic metres. On the other hand, in the time frame under consideration, Taiwan too will export over 0.5 million cubic metres of softwood lumber by the next century (APRENet, 1997). However, these countries are likely to be more significant in the competition for plywood.

IIASA, CINTRAFOR and FAO forecasts

The International Institute for Applied Systems Analysis (IIASA) is working on a project which involves, among other things, the forecasting of timber output from Russia and it is worth reviewing those forecasts here. Their forecasts follow, to a certain extent, the factors outlined above.

The major factor which IIASA identifies as influencing export levels is the increase in rate of domestic demand implying that foreign exchange earnings are not seen as being paramount and this is in line with Backman's earlier comments. IIASA forecasts of net trade in forest products depend on the level of growth of Russian GDP in the future. The results of three of their five scenarios are illustrated in Fig. 6.5. The figures in the diagram refer to East Siberia and the Far East. Thus, they are not directly comparable with those quoted earlier which related mainly to the Far East only. The differences in the scenarios relate to the assumed growth rate of the Russian economy during the study period and these are given in the legend to Fig. 6.5. The effect of the growing internal demand in diverting the timber harvest to the domestic market is immediately apparent. Nilsson (1997a) argues that the most likely outcome is the middle case, where the Russian economy grows annually by 5% for the first part of the forecasting period and then by 2.5% for the final 25 years. This he refers to as the 'modified high growth scenario' and in this case exports dry up around 2015.

The forecasts in Fig. 6.5 assumed a constant manufacturing capacity and

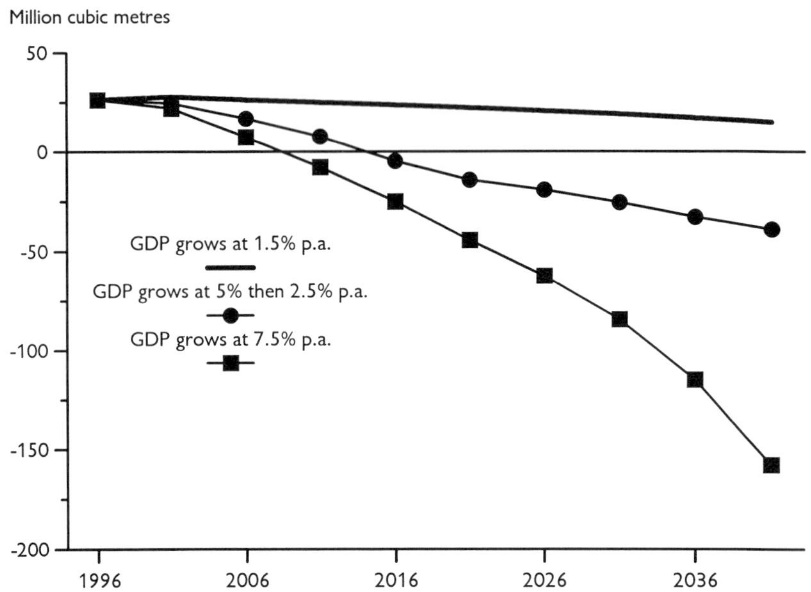

Fig. 6.5. IIASA forecasts of exports from Eastern Russia

fibre availability. If these change then, clearly, the forecasts themselves would alter. To investigate this, the effect of different levels of investment on the export potential of Eastern Russia were considered. The implications of these on the modified high growth scenario are shown in Fig. 6.6. In Case I investment allows productive capacity to increase by 10%, in Case II by 25% and in Case III by 50%.

According to Case III, the potential of Eastern Russia to export appears to be immense, reaching more than 180 million cubic metres a year in the final part of the forecasting period, a figure close to the present total AAC for Asian Russia. However, the level of investment required to bring this forecast about is equally impressive, more than US$40 billion per 5-year period at the beginning of the forecasting period, rising to over US$100 billion at the end. Compared with these figures, recent Japanese investments in the Khabarovsk region have fallen from nearly US$15 million in 1991 to only US$42,000 in 1997 (Kondrashev, 1998).

The potential to add to the capital inflows required is something which, in part at least, will hinge on political factors. Relations between Japan and Russia are soured by the question of the sovereignty of the Sakhalin islands. In a statement on Japan's policy to the Russian Federation a Ministry of Foreign Affairs press release began with the basic objects of resolving 'the Northern Territories issue, thereby concluding a peace treaty and fully normalizing relations ... '. They further state rather pessimistically that '[d]ue to continuing non-payment of uninsured trade debts and inadequate legal and taxation systems

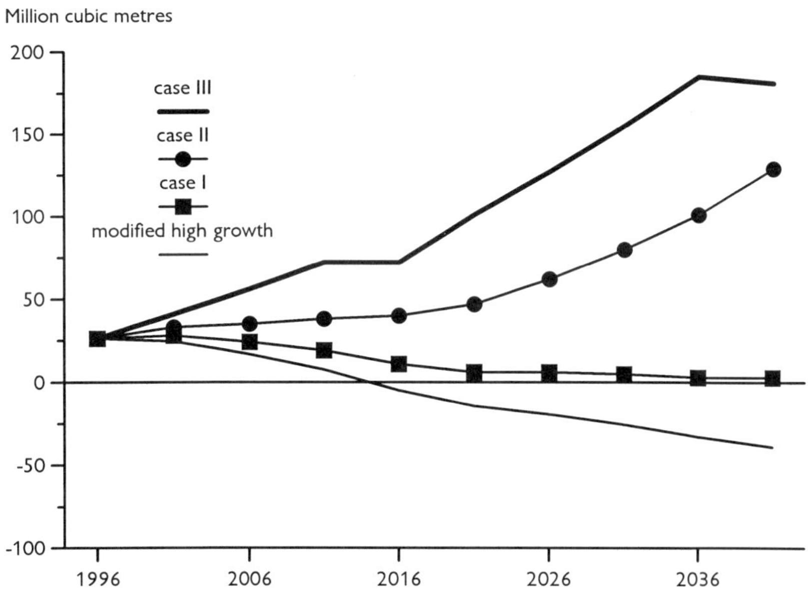

Fig. 6.6. Alternative forecasts of exports from Eastern Russia

in Russia, there is little economic activity between Japan and Russia' (MOFA, 1997).

CINTRAFOR has also produced a forecast for the period until the year 2000 (Backman and Waggener, 1994). This suggests that the annual output from Russia as a whole will be around 250 million cubic metres in the period 1990 to 1995 of which 24% is felled in the Russian east (defined again as Eastern Siberia and the Far East). In the period after that, until 2000, they produce forecasts for two scenarios. In the optimistic scenario the total delivered supply was projected to reach 338 million cubic metres. In this case the Russian GDP was projected to grow which allowed a growth of investment in the timber industries. The output in the Russian East was forecast to be just over 84 million cubic metres. Under this case exports to the Pacific Rim countries were suggested to be around 12 million cubic metres a year.

Under a pessimistic scenario they argued that domestic demand would absorb the total output of timber but they suggested that their analysis indicates that the central authorities would sacrifice domestic consumption to maintain exports at roughly their present levels. Exports to the Pacific Rim were thought to be in the region of 5 million cubic metres.

The CINTRAFOR forecasts looked at the effect of price reforms in Russia. The present prices and exchange rates fail to reflect the international situation and this, they suggest, is leading to a misallocation of resources within the country. Changes in relative prices and costs will clearly affect the accessible forest and, therefore, the potential allowable cut. It will also predispose the firms operating in Russia to the export of either logs or sawnwood and other forms of processed timber. They showed that a full price reform, bringing prices in line with world levels, would increase costs somewhat and so reduce timber output to levels near the pessimistic scenario. However, a shift towards the Russian East was observed in this case as moves towards more accessible timber stocks were stimulated.

Forecasts from the FAO regional trade report (Zhang *et al.*, 1997) show exports from Russia increasing by about 18% over the period 1992 to 2010. The share of this going to Japan increases and Japanese imports rise by 27% to about 5.3 million cubic metres. Competition from Korea or China does not materialize as their imports stagnate at present levels. The export of coniferous sawnwood from Russia decreases over the study period and Japanese imports decline with them. Also, despite the use of larch in plywood manufacture, the report by Zhang *et al.* shows a decline in plywood production in Russia and a corresponding decline in its exports. So, in the same way that their methodology seemed to favour log production in the case of North America, their forecasts for Russia also show a bias towards log exports. Given the nature of the forecasting method it is possible that this is a manifestation of profit maximizing behaviour over revenue or foreign exchange maximization. However, unlike their forecasts for North America and Southeast Asia, these predictions do at least follow the present trends and would be compatible with a slow growth rate of the Russian economy and low inward investment.

Concluding remarks

The present levels of imports into Japan are bracketed by the forecasts of CINTRAFOR, which suggests that if their pessimistic forecasts are correct, Japan could have difficulties in continuing to attract imports. The conclusion to CINTRAFOR's forecasts suggests that their optimistic scenario is the more likely but that without substantial capital inflows further capacity expansion is unlikely if not impossible. They suggest that the trade of forest products will continue at levels recorded in the 1980s and this seems feasible if the optimistic scenario is the correct one.

The forecast which IIASA considers most likely, the modified high growth scenario, indicates a similar problem for Japan in the medium term as the growth in the domestic economy reduces the timber available for export. It is not clear what effects of the price of timber are incorporated in their model. If the modified high growth scenario turns out to be correct, it is likely that the drying up of exports would be countered with price increases which would increase the economic availability of timber and allow exports to the Pacific Rim and Japan to continue.

The reduction in output in recent years in the Far East means that the optimistic CINTRAFOR figures for exports are now greater than the Far East's total output. In one sense, the spare capacity in the region makes forecasts of future increases more likely and so export levels could be maintained for the near term at least. After that it is clear that investment levels and internal demand constraints will be of crucial importance. Not only is the investment level critical for output levels but it also determines the form of that output. The FAO figures of Zhang *et al.* (1997) are, unlike their forecasts for North America, more easily believable and would correspond with the continuing stagnation of the Russian economy. However, the cost levels in Russia suggest that increased output of sawnwood and panel products would, if the quality were high enough, find a ready market in Japan.

New Zealand and Chile

Unlike the other countries supplying Japan with timber, New Zealand and Chile are different in that the wood they are exporting comes largely from man-made forests. There are relatively few ecological constraints on the felling of this timber unlike the situation in the tropics, North America and Russia. It has been planted specifically for industrial use and the problems of falling management intensity seen in Chapter three with regard to the Japanese timber estate do not arise.

A major similarity of New Zealand and Chile is in the species supplied, radiata pine (*Pinus radiata*). It accounts for the bulk of the commercial forests in the two countries and means that they are usually thought of together by the Japanese. Radiata pine is a timber which is easily dried and treated and can be worked to give a fine finish. It is stable and structurally strong with good nail and screw retention. These features mean that it would be ideal for construction work in Japan but the timber around knots may tear under processing and the knotty nature of a lot of the timber along with a perceived lack of strength means that it is regarded very largely as a species for low-grade end-uses, being used to a great extent for packaging. Thus, the demand is dependent to a large extent on Japan's export performance as the main area of utilization is in the crating of items such as cars, large electrical equipment, etc.

Radiata pine, which is a native of California, is particularly well-suited to the conditions in New Zealand where it exhibits an average annual growth rate of around 25 cubic metres and is ready to fell at the age of 25 or 30 years. This means that the forests planted in the 1970s will be mature at the turn of the century. Growth rates in Chile, in the central-southern regions where radiata pine is usually planted, average around 28 cubic metres a year which is greater than the figures recorded in New Zealand. This gives a rotation length of between 20 and 30 years, somewhat shorter than that in New Zealand.

New Zealand has been supplying Japan with timber since the late 1950s, in general as logs. At the moment, New Zealand holds a market share of just over 3% in RWE. Chile's market share is 3.8% and this greater figure is a reflection of the extra imports of lumber from Chile rather than logs. In 1997, for example, the log exports from New Zealand to Japan were just under 2 million cubic metres whilst those from Chile were only 183 thousand. On the other hand, the imports of lumber were only 338 thousand cubic metres from New Zealand but 616 thousand from Chile (Araya, 1998).

New Zealand and Chile's forest resources

The forest cover of New Zealand is a little under 8 million hectares which is around 29% of the total land area. The area of natural forests stands at 6.4 million hectares whilst man-made forests cover just under 1.5 million (Brown, 1997). However, in production terms, man-made forests are by far the more important. Most of the natural forests in State hands form national parks and those in private hands are set aside for protection purposes. The result is that natural forests account for only 1% of timber output (White, 1997).

The area of commercial or plantation forest increased by over 50,000 hectares a year during the 1990s and most of this expansion has been undertaken by small-scale owners and farmers (Horgan and Maplesden, 1995). This man-made forest is almost entirely coniferous. Of the total area of man-made forests of 1478 thousand hectares only 41 thousand is non-coniferous. By far the most important species in the man-made forests is radiata pine. In fact, New Zealand holds about one-third of the world's radiata pine resources and the species accounts for 93% of the commercial forests and 95% of new planting. The man-made forests of New Zealand are, as a result, over 90% radiata pine with the second most planted species, being Douglas fir. Hardwoods make up only about 3% of the total area.

The commercial forests of New Zealand are comparatively young, with an average age of 13.3 years compared with a rotation length of between 25 and 30 years (Brown, 1997). The age class structure is such that a large increase in supply is predicted as maturing timber stocks come on-stream.

The present nature of New Zealand's forests is the result of a major planting programme which started between the World Wars. A Royal Commission report in 1913 predicted a timber shortage in the country because of the felling of the natural forest. The government mounted an afforestation programme and there was a planting 'boom' in the 1930s which reached a peak with nearly 40,000 hectares a year being planted. By 1936 plantation forests of over 300,000 hectares had been established, mainly of radiata pine.

During the Second World War planting was negligible but afterwards planting began to pick up again. However, it was not until the 1970s that a second 'boom' occurred with planting exceeding that undertaken in the years between the World Wars. This boom was more widespread than the inter-war one and man-made forests were established in most regions of the country. This was brought to an end by the economic restructuring which the New Zealand government undertook in the late 1980s. Recently, planting has again accelerated. Unlike the previous plantings, which were more-or-less evenly divided between government and private land owners, the present planting is entirely by the private sector.

In the early years, radiata pine was not particularly favoured in New Zealand's domestic market, being used mainly for boxes and concrete forming. It was sold ungraded and 'was regarded almost universally as an inferior species because there was no shortage of high-quality indigenous timber' (Horgan and Maplesden, 1995). However, as a result of the government's efforts, radiata pine has been accepted in the domestic market. In 1947 Forest Service graded lumber was permitted as one of the framing timbers for housing. From there the timber moved to be used in home building and furniture making where previously native timbers had been predominant. Despite the lack of export orientation in the early days, the result now is that the forest industries have become a major part of New Zealand's export profile.

The forest area of Chile is 22.6 million hectares of which 13.7 million are natural parks and forest reserves. The area of commercial forest in Chile is nearly 9 million hectares, most of which is native hardwoods. The man-made forests cover only 1.3 million hectares but are predominantly radiata pine, with the other species, mainly eucalyptus, making up only 13% (Jélvez *et al.*, 1990). The result is that Chile accounts for over one-third of the world's radiata pine forests. The plantations of radiata pine are part of an explicit government policy to encourage non-traditional exports at the expense of the previous export staples of mining resources (Gwynne, 1993). Like New Zealand, the proportion of total exports accounted for by forest products has risen, from just over 3% in 1973 to over 10% now.

In the early 1970s the Chilean government began a policy of opening the country to international trade with a reduction of tariffs and the removal of non-tariff barriers. The result was that there was an increase in non-traditional exports. This paved the way for increased investment in radiata pine.

Although the planting of radiata pine began in the early twentieth century it was not until the 1970s that the areas planted began to increase noticeably reaching a peak of nearly 100,000 hectares in 1976. Since then plantings have settled down to average around 65 thousand hectares a year. Thus, like New Zealand, the radiata pine forests of Chile are relatively young and an increase in supply of timber can confidently be expected in the future.

Another change, which is likely to occur in the future, is the nature of the raw material being harvested. Maturing radiata pine forests have largely been managed using non-intensive techniques and yield small-diameter knotty logs which are mainly sold for pulpwood and low-grade sawlogs. Recently, intensive management techniques, which involve regular pruning and commercial thinnings to produce higher quality knot-free sawlogs, have been introduced.

With the non-interventionist policies went, paradoxically, sector-specific policies aimed at forestry. Under a law passed in 1974, the state intended to return order to the ownership of land in the forestry sector by establishing a dominant role for the private sector. Grants were established giving up to 75% of the costs of establishment of plantations and establishing a tax-free environment for forestry operations. The main reason was, according to Gwynne (1993), the close relationship between the military government and the forestry sector. The result was that, compared with the 1974 figures when state enterprises were responsible for around half of all plantings, by the end of the 1970s almost all plantings were in the private sector. In 1984 pruning subsidies were introduced and this accounts for the increase in pruning activity noted earlier.

Timber output and Japanese imports

New Zealand

In the 1940s and 1950s there was little export orientation to New Zealand's forest industries, the emphasis being on developing an internal market for the exotics that were being planted. The log trade with Japan began as an experiment by the Japanese trading houses to secure a supply of timber as the output from Japanese forests began to fall. This was particularly true of the lower grade Japanese species. Apparently, at this time, although Japan accounted for over 90% of New Zealand's log exports, this market was seen mainly as a means of using timber that was surplus to domestic requirements. Indeed it was viewed as 'the preserve of unwanted, damaged and low-grade material' (Horgan and Maplesden, 1995). Most logs apparently suffered from sapstain by the time they reached their export markets, a feature which immediately reduces their possible uses in the Japanese markets.

However, in the mid-1980s the situation began to change. With the privatization of the forest industry, the need for liquidity in the private sector and the production orientation of the purchasers of forests became apparent and exports rose rapidly.

This is shown in Fig. 6.7 which uses FAO figures. The recent acceleration in production is apparent and results from the planting undertaken about 30 years ago. Another feature which is clear from the diagram is the increasing levels of exports of industrial roundwood. One of the lines shows the log exports as a percentage of total production and this has also increased in recent years reflecting the words of Brown (1997) who states that 'the higher harvest potential of New Zealand's forests has not been matched by development of processing capacity'. In many ways the increase in exports seems to have taken New Zealand by surprise as it was only in the second planting boom that export potential seems to have been explicitly considered. However, the effect has been that, now, the forest sector accounts for around 15% of all export earnings.

In the early years log exports were almost entirely destined for Japan but in the mid-1980s Korea and, to a lesser extent, other countries entered the market. As a result the proportion log exports destined for Japan fell from nearly 100% to

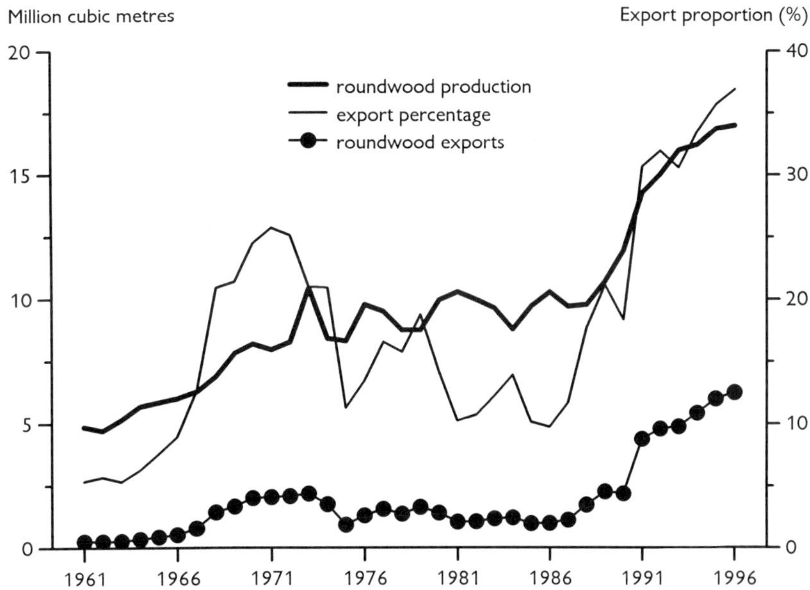

Fig. 6.7. New Zealand roundwood production and exports

less than half of New Zealand's log exports. Provisional figures for the destination of New Zealand's exports for the year 1996 are shown in Table 6.4. The table shows the importance of Japan to the export trade in New Zealand. However, it also shows the competition faced by Japan from Korea.

The level of Japanese imports of logs and lumber is shown in Fig. 6.8. Two features are immediately apparent; the sudden drop and then increase in log imports and the relatively low level of lumber imports. The first of these features was the result of a relative scarcity of mature timber in New Zealand. Post-war

Table 6.4. New Zealand's timber exports (thousands of cubic metres)

	Logs	Lumber	Fibreboard	Plywood	Other panels
Australia		293	4	24	15
China	81	4	81		1
Korea	3086	89	9	3	2
Philippines	136	15	6		1
Taiwan	127	79	37	9	6
North America	54	136		1	
Other	53	76	40	11	14
Japan	2103	269	177	59	88
Total	5640	961	354	107	127

source: White (1997)

Fig. 6.8. New Zealand log and lumber imports

plantings were yet to reach maturity and there was a hiatus in supply as the plantings from the inter-war years and other forest sources were used or set aside for environmental uses. During this period radiata pine imports from Chile began.

The relatively low level of lumber imports shows again the propensity of Japan to import raw materials rather than finished goods. One aspect of the trade with Japan is the export of squares rather than logs. This allows a higher level of recovery from the imported squares than would be the case is logs were imported. The volume of squares imported from New Zealand has been significant, occasionally accounting for almost half the imports. The mills using radiata pine tend to be larger than average and so the import of logs and squares could well continue for economic reasons (Maplesden, 1993). However, log exports are a somewhat sensitive issue, being seen as the export of jobs. Brown (1997) suggests that companies might limit log exports to avoid government intervention if the prevailing state policies move 'leftward', away from free markets towards a slightly more interventionist position.

Chile

Output of industrial roundwood in Chile has accelerated rapidly in recent years and this is shown in Fig. 6.9. Unlike New Zealand, the proportion of timber production coming from man-made forests has been decreasing over time and now stands at around 75% of the total (Yanasachi, 1997). Like New Zealand

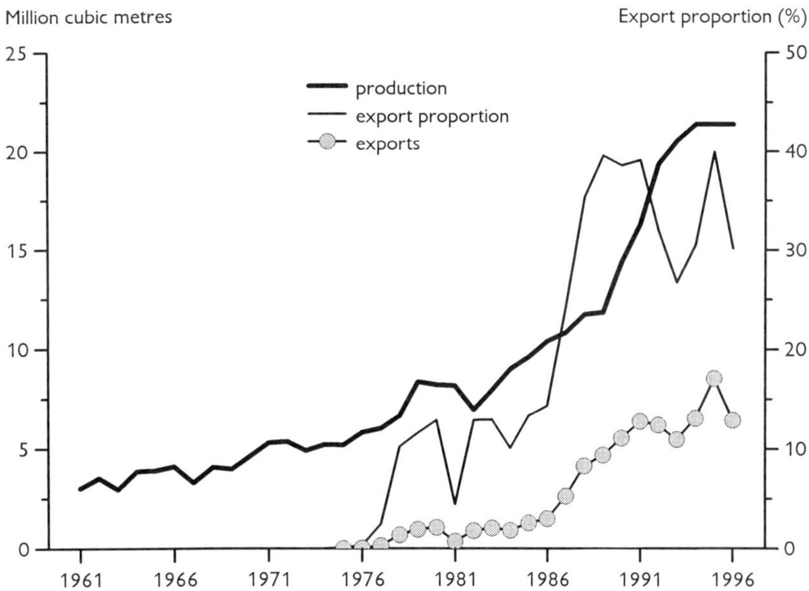

Fig. 6.9. Chile's roundwood production and exports

however, the success of the export policies can be seen in the increase in the proportion of the output that has been exported.

One major difference between Chile and New Zealand is the form in which the timber products are exported to Japan. Whilst the proportion of roundwood from New Zealand is high, imports from Chile are mainly sawnwood. Over recent years, against a rising level of lumber imports, the imports of logs have declined. In 1985 the proportion of log and lumber imports accounted for by logs was 79% in the case of Chile and only 65% in the case of New Zealand. By 1997, however, the log imports from Chile accounted for under a quarter of the total log and lumber imports whilst the New Zealand figure had risen to over 85% in unconverted units (Araya, 1998).

Chile also serves a more diversified export market than does New Zealand with forest products being shipped to Europe, Asia and South America in very roughly the same quantities. However, Japanese trade dominates the country's exports to Asia with Korea and China again being Japan's main competitors.

Future prospects for imports from New Zealand and Chile

With an age class structure reminiscent of that of man-made forests in Japan, it is clear that there is an enormous potential for increased production from New Zealand's forests and, therefore, the supply side of imports into Japan should pose no real problem. Unlike the Japanese case, where the economics of private forest production mean that a lot of the maturing timber is unlikely to be felled, there are no real reasons why the maturing stock of exotic conifers in New Zealand will not be felled. The typical internal rate of return of a radiata pine forest is in the region of 7-8% and, given the relatively short rotation lengths involved, it is in forestry terms a fairly safe and profitable investment.

Figures for the potential output of New Zealand's forests depend on the assumed rotation length and the area of land which is to be planted in the future. In Fig. 6.10 forecasts of the productive potential of New Zealand's forests are presented and the increase in output which is anticipated is clear. The figures are based on those produced by the New Zealand government and are adapted from those given in Konohira (1995), Brown (1997) and Yanasachi (1997). The potential output of the current area of man-made forests are given for three proposed rotation lengths. The line labelled 'normal' shows the output that will occur if felling is undertaken at the normal rotation age of around 30 years. The short rotation used is 25 years and the long rotation is 35 years. As would be expected, the longer rotation length gives a greater volume of fellings in the later years but the acceleration in output occurs later in time. It is argued that the 'normal' scenario represents the best forecast of output from the present forest estate with the other two cases representing upper and lower bounds to the likely output. In all cases, increased growth rates from genetically improved stock have not been considered and so the forecasts, in this sense, are thought to be

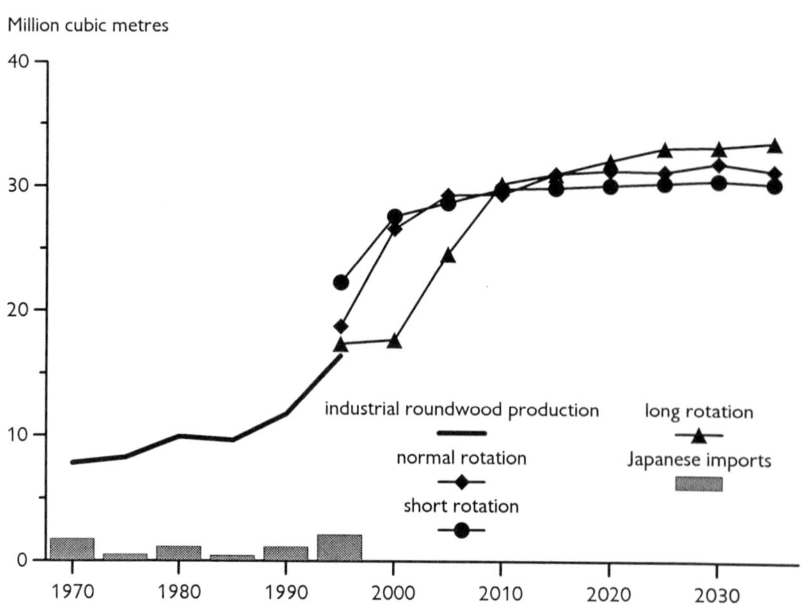

Fig. 6.10. Forecasts of New Zealand production

conservative (Brown, 1997).

Another factor which is likely to influence the Japanese market is the pruning regime undertaken. A typical silvicultural regime for radiata pine consists of three prunings before it reachs age 10. This gives a clear 6 metre butt log which fetches a premium for use by furniture manufacturers or plywood producers. This silvicultural system has developed over recent years. Figures in Horgan and Maplesden (1995) show that about 90% of the forests aged 36 and over have not been pruned whereas in the youngest forests around two-thirds have been pruned. As a result, the forecasts above indicate that pruned log output will increase nearly three times in the period to 2010, reaching 4.2 million cubic metres.

However, not all are happy with this and the pruning debate is a current one in New Zealand forestry. Recently, with reduced timber prices and tightening cash flows, the advisability of pruning has been much questioned. Pruned timber occasionally exhibits resin pockets and inherent stresses. This is of significance for the Japanese market as such defects would preclude the use of the timber in many areas. However, it is notable that some Japanese companies are pruning four times to greater heights. One of the criticisms of radiata pine for construction is that the branch whorls reduce the strength of the timber when used in the long spans needed for traditional building (Schäffler, 1998).

Figure 6.11 shows forecasts which have been produced on the assumption that land is planted and new commercial, man-made forests are initiated. With the

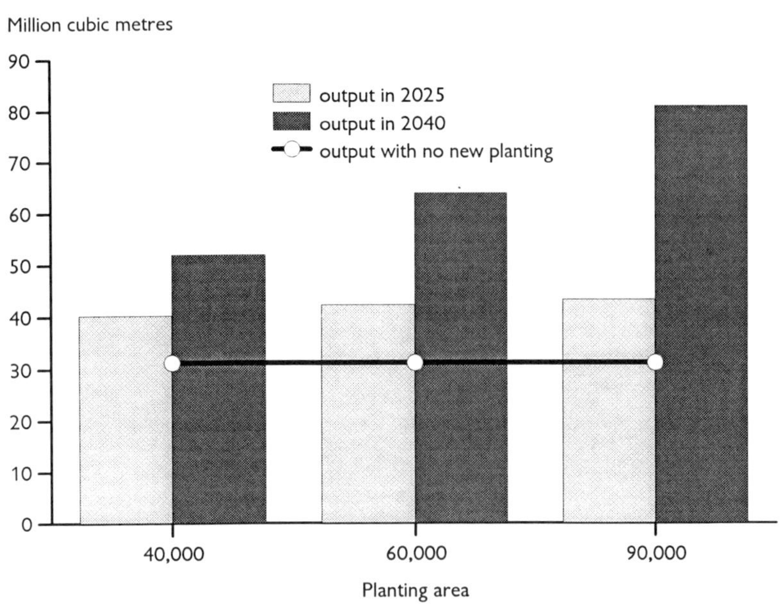

Fig. 6.11. The effects of planting on forecasts

shortness of the rotation length in New Zealand this will have an effect which will be felt relatively quickly. Output levels for three different future planting regimes, an extra 40,000 hectares a year, an extra 60,000 or an extra 90,000 are shown. The output figure for 2025 with no new planting is taken from Fig. 6.10 and is reproduced for reference. The output levels in 2025 are not dissimilar from those for the 'no new planting' scenario but in the later years the extra planting has an effect and output increases. The New Zealand Forest Research Institute has produced estimates that suggests that the most likely planting activity in the future will lie around the 40,000 hectare levels (quoted in Brown, 1997).

Chilean projections of their output potential show a very similar pattern to that for New Zealand. In their models, assuming a continued planting rate of 57,000 hectares a year, a figure which is close to the present average, and using a 27 year rotation, output is expected to reach 27 million cubic metres a year by the turn of the century (Jélvez *et al*., 1990). The harvested volumes are expected to be roughly equally divided between sawlogs and pulpwood which is somewhat different from the present proportion which shows a greater percentage of pulpwood.

In the longer term large increases in output are anticipated. Figures reported in Yanasachi (1997) show that output is expected to increase to 56 million cubic metres by 2020 and, in this total, radiata pine will contribute 47 million. Eucalyptus production will also increase but the main feature is the continuing increase in sawlogs rather than pulpwood in the mix. At the moment the proportion of pulpwood in the radiata harvest is 44% but this is projected to drop to under a third in 2020.

The figures in the FAO regional report by Zhang *et al*. (1997) show that exports of coniferous logs from New Zealand are expected almost to quadruple by the year 2010 while from Chile the increase is nearly tenfold. The share of these exports being taken by Japan also increases. By the year 2010 Japan is expected to take about 86% of the exports from New Zealand and from Chile the figure is 89%. Together this represents 24 million cubic metres of logs, nearly ten times the present figure. Clearly, for such figures to be realized the present perception of radiata pine as an inferior species will have to be altered. Surprisingly given the present propensity of Chile to export sawnwood rather than logs, the FAO forecasts for coniferous sawnwood exports show Chile's exports decreasing by about a half and New Zealand's remaining more-or-less stable.

The main problem faced by both New Zealand and Chile in exporting to Japan seems to be the perceptions of the product and this casts doubt on the FAO forecasts, as the authors of that report mention. In a recent address to the Japan-New Zealand Business Council, Mr Shoichi Tanaka, of Nissho Iwai Corporation stated:

> '... we cannot but conclude that radiata pine does have some disadvantages, with drawbacks that include bluestain, young growth, and

> lack of strength. To overcome those disadvantages, a key point will be price competitiveness. The marketing policy for radiata pine for use in plywood - keeping a close watch on export prices for Russian larch shipped to Japan - will be crucial. Frankly speaking, radiata pine logs and green lumber are not highly estimated. To cope with any substantial growth of supply in the future, New Zealand will have to produce wood products that are high in added value.' (Tanaka, 1997a).

One of the problems identified by Tanaka is the competition of larch for the softwood plywood market from Russia. This timber, he argues, is of very high quality, with narrow annual rings, small knots and a fine grain, all of which are elements on which radiata pine cannot really compete.

Competing on price is not something which Chile and New Zealand would be particularly interested in although it would be one way to avoid the substitution of other packaging materials in the present market for radiata pine (Kawata, 1992). However, the future, according to Tanaka, seems to lie in plywood and other panel products to compete in the prefabricated housing market. Also, Tanaka seems to suggest that a radical change in policy is needed, as he argues that planting Douglas fir might be a policy worth following. Despite its longer rotation length, it commands a higher price and can be targeted at the more valuable markets for traditional housing.

Concluding remarks

The figures show that, on the supply side, the harvests from New Zealand and Chile are likely to be able to meet any potential demand from Japan in the foreseeable future. Indeed it seems to be the case that marketing the output is more likely to be the problem for these countries. The trends in the housing markets favour the output of the timber from these countries and the areas of housing in which the traditional species find favour are diminishing in importance. However, the prices paid by Japan for American timber and sawnwood and the move to Europe to secure supplies all suggest that radiata pine has some way to go to be accepted as a high quality timber in the Japanese market.

Imports from Europe

To conclude this chapter a brief look at the imports of forest products from Europe is in order. The imports from this region have only just become large enough to warrant a special section in the usual sets of Japanese statistics produced in the Forestry White Papers.

Some log imports from Europe occur and these are mainly from Finland which, in 1997 exported just over 100 thousand cubic metres of logs to Japan.

European log imports account for about 0.7% of the total log imports into Japan and only just over 1% if *nanyouzai* is ignored. A more significant field for European countries, and especially Scandinavia, is sawnwood. Japanese imports of sawnwood from Europe are shown in Table 6.5.

Table 6.5. Imports of sawnwood from Europe

	1992	1993	1994	1995	1996	1997
Finland	1,514	128,176	214,648	365,717	448,090	558,081
Sweden	1,188	82,464	204,316	209,327	349,786	839,186
Norway	162	8,556	25,126	31,000	55,237	47,646
Austria	8	4,909	77,590	157,330	285,019	465,116
Germany	1,910	9,000	37,874	55,042	35,084	97,085
Others	5,423	27,254	31,382	47,743	61,522	101,773
Total	10,205	260,359	590,936	866,159	1,234,738	2,108,887

All figures in cubic metres.
source: Araya (1997; 1998)

Two features emerge from the table. First, there has been a dramatic increase in the imports of sawnwood from Europe over recent years and most of this has come from Scandinavia and Austria. However, secondly, the overall level of imports are still fairly small compared with the total sawnwood imports of Japan of 12.5 million cubic metres. However, the proportion of European sawnwood in the overall Japanese market has been steadily increasing, from barely 0.1% in 1992 to around 17% in 1997. Also, these imports are beginning to represent significant markets for the exporting countries concerned. Using FAO figures for sawnwood exports, the Japanese market now takes around 8% of the exports of Finland and Sweden.

Part of the reason for this increase was the sudden increase in the price of North American sawnwood imports in 1993. The price of 10.5 cm pillars rose by over 10% on the previous year's level. The Forestry White Papers comment on the thoroughness of the quality control of timber from the area and the price advantage of the timber.

As well as sawnwood the European producers are making their presence felt in the market for glued laminated timber. The overall imports of this in 1997 ran at 289,377 cubic metres of which the largest single suppliers were the Americans. However, the market share from European suppliers was 37%, the major suppliers being Sweden, Germany and Finland. This, again, is a relatively new phenomenon, with the corresponding market share in 1992 being less than 0.5% (Araya, 1998). It seems that the Japanese have followed a fairly typical strategy, for them, of first importing small quantities of forest products to establish links and assess quality and then, if satisfied, increasing their imports.

The impression gained is that the imports of processed forest products from Europe can compete with American imports and so can gain a significant market share if the price is suitable. The situation, therefore, seems to depend on the output of products from North America, with Europe positioned to fill any short fall that the environmental moves cause in the USA and Canada. The movements to preserve old-growth or to reduce timber output for environmental reasons, although present in Europe, do not seem to be of the same intensity as in North America. Similarly, however, the capacity to expand output in the short run is not really present and so the demand for forest products from Europe will serve to increase competition in that area and, possibly, drive prices up.

Conclusions

The look at these countries supplying Japan has shown that the Japanese market appears to be one which is, paradoxically, at the same time resisting the changes in the markets outlined here by bidding for the remaining harvests from the West Coast of the USA, as was seen in the previous chapter, and expanding and diversifying its sources of supply. The forests of Russia, whilst having been witness to a significant drop in output have not ceased to be an important source of coniferous imports for Russia and so it is unlikely that any recovery in output in the near future would allow a significant increase in exports. The ability to compete in the area of processed forest products will depend upon the ability to attract capital investment. New Zealand and Chile will both be able to expand their exports but the limiting factor here seems to be a marketing one, the need is to expand the market in Japan.

Chapter seven:

A Review of Forecasts

Introduction

The arguments so far have shown that Japan is an important country in the international trade in forest products and that many of the issues which dominate world forestry, such as the management of tropical forests, the development of the Russian forestry sector, North American environmental movements, have an influence on, and are influenced by, the Japanese market. It follows that forecasts of the forestry sector in Japan are of interest, not to say importance, to many forestry agencies and forest owners around the world.

This chapter will review the forecasts which have been produced concerning the Japanese forestry sector by the Ministry of Agriculture, Forestry and Fisheries (MAFF) in Japan. The latest one has only recently been published and a summary of it is also available in the FAO Asia-Pacific Sector Outlook Study (FAO, 1998).

The 'Long Range Prospects'

In 1951 the first 'Basic Forest Law', which established a framework for the planning and control of Japan's private and national forests, was passed. It has since been amended a number of times but, for the purposes here, the 1964 amendment is of importance as this modification saw the inclusion of Article 10 which required the Japanese Government to produce a 'Basic Forest Plan' and 'Long Range Projection of Demand and Supply of Important Forest Products' (*Shinrin keikaku seidou kenkyuukai*, 1992). The latter is essentially a long-term forecast of the likely outcomes in the forestry sector if the policy elements of the Basic Forest Plan are successfully implemented. The Long Range Prospects (LRP), as the Japanese often translate them, are produced at irregular intervals, whenever the necessity arises. This is usually as a result of changes in the condition of the forest resources, the market for the important forest products or other economic circumstances.

The first LRP were produced in 1966, they were revised in 1973 and 1980 as a result of the oil shocks and then again in 1987. The latest version was published in 1997. The first of these LRP were based on simple judgmental extrapolations but recently they have been the result of a major forecasting effort. The LRP of 1987 (*Shinrin keikaku kenkyuukai*, 1987) were based on a mixture of econometric and simulation techniques and were presented in a 414 page book. The latest set of forecasts has been produced using the same basic modelling methods and this time the forecasts are presented in a 416 page book (*Shinrin kihon keikaku kenkyuukai*, 1997). Unless otherwise attributed, all of the forecasts in this chapter originate from this source.

A feature of all of the forecasts has been their overestimate of the output from Japan's domestic forests and, therefore, an underestimate of the imports of forest products into Japan. However, before examining these forecasts in detail it is worth looking a bit more carefully at the nature of the forecasts and forecasting in Japan in general.

Forecasting in Japan

The wording of the Basic Forest Law refers to the LRP as *mitooshi* which is usually translated as 'prospects' or 'outlook' and the Basic Plan as *keikaku* which is usually translated as 'plan'. However, English language versions of Japanese publications often translate the word *mitooshi* as 'projections' and the word 'forecast' crops up in the discussion of the figures (see for example MAFF, 1980). The nature of the words used is of some importance as plans and forecasts tend to have slightly different implications in Japan. To quote at length from a book by Nakamura (1981):

> 'Japanese plans possess three basic characteristics. First, they indicate the desired direction of economic and social development; second they indicate the policy direction the government should take in order to achieve these ends; and third, they indicate behavior guidelines for the people and the businesses. ..... on the whole, the plan figures are somewhere between predictions and guidelines.'

Nakamura is mainly talking about the main macro-economic *keikaku* and *mitooshi* which were drawn up during the years of rapid economic growth and after, but the last sentence in the quotation is useful here as it suggests that the forecasts published by the Japanese should be treated with a little caution. There have no doubt been attempts to 'talk up' the market and to attempt to encourage domestic private investment on the back of projected increases in domestic output. Some Japanese admit that the result is that, now, the LRP are not taken particularly seriously. However, the forecasts have been reported by others in a fairly direct way as if they represent the best possible outcome of a disinterested forecasting effort, not least by the present writer (Blandon, 1996). Thus, the overestimates in the output of domestic logs in the LRP do not necessarily represent careless forecasting but rather a divergence of what the government would like to see and what actually transpired.

The 1973 and 1980 long range prospects

The early forecasts were fairly basic in nature and only attempted to predict the output for broad product groups. They can be summarized briefly, the point being to illustrate the optimistic nature of the forecasts for the self-sufficiency ratio and domestic output.

Figure 7.1 shows the projected levels of domestic production under the 1973 and 1980 LRP and the optimistic production levels are clear. Both of these forecasts contained only gently increasing levels of overall demand, so the increased domestic production translated more-or-less directly into lower levels of imports. The relatively stable demand levels were the result of the perceived effects of the oil shock and the structural shift in the Japanese economy from a period of rapid economic growth to one of moderate growth.

In the 1973 projections total demand was forecast to rise to 135 million cubic metres RWE in 1981 and then to 147 cubic metres in 1991 giving a

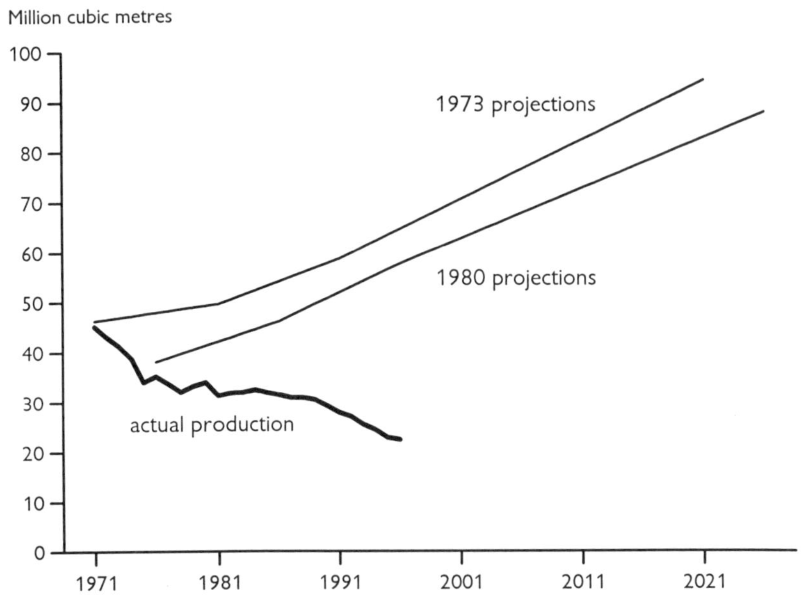

Fig. 7.1. Domestic production in the early forecasts

projected self-sufficiency ratio of just under 40%. However, the increase in domestic production in the later part of the forecasting period gave rise to a self-sufficiency ratio as high as 62% in 2021. The 1980 LRP again predicted a rapidly increasing domestic output but the level of demand was forecast to rise less slowly. Even so, the self-sufficiency ratio was predicted to be 43.3% in 1996.

Despite these forecasts appearing optimistic in retrospect, the explanations at the time made them seem quite reasonable. For example, the accompanying English language commentary to the 1980 projections argued that the movement to second growth cutting in America, the reduction of supply from tropical countries and the switch to more remote felling sites in Russia would mean that:

> '[t]he ratio of domestic supply to total demand for timber in Japan will rise at that time when the supply of foreign timber begins to level off. The supply of domestic timber will increase gradually as the cutting period for young man-made domestic forest arrives.' (MAFF, 1980)

Such an argument should not be too criticized. Despite the aspect of encouragement contained in Japanese government forecasts, the translation of the increasing maturity of timber into at least a small increase in domestic supply would have been difficult to avoid. This is especially true if the magnitude of the maturing timber stock shown in Chapter one is considered.

However, not only were the forecasts of domestic supply an overestimate, the Japanese government also failed to foresee the turndown in domestic demand which was to occur. The forecasts of the 1980 LRP broke down overall demand into a number of sub-categories and each of these registered an overestimate. This is shown in Table 7.1.

Table 7.1. The accuracy of the demand forecasts in the 1980 Long Range Prospects (LRP)

	1986 forecast	1986 actual	1996 forecast	1996 actual
Sawnwood	62.6	44.9	65.4	49.8
Pulpwood	36.1	33.6	44.9	43.8
Panels	14.9	10.9	17.6	15.7
Shiitake bed logs	3.6	1.9	4.1	1.0
Other	1.2	5.1	1.2	3.7
Total	118.4	96.4	133.2	114.0

All figures in millions of cubic metres RWE.
source: MAFF (1980) and various Forestry White Papers

What is apparent is the optimism in all of the figures, even those for *shiitake* bed log production which, if anything, the Japanese should have been able to forecast accurately. These overestimates are all the more surprising when it is considered that the premise underlying the revision of the LRP was that Japan's economy was entering a period of slower growth rather than the rapid growth it had seen before. The figures also seem to imply that there was an unpredicted movement away from the use of solid timber in the Japanese markets towards the use of panel products. This is consistent with the previous discussion and the trends in housing starts discussed in Chapter two.

The 1987 LRP - the establishment of a major forecasting model

The 1987 LRP represented the result of a concerted forecasting effort based on a sophisticated model of the Japanese timber market. This model was mostly reported in Japanese with the equations described in detail in the major publication (*Shinrin keikaku kenkyuukai*, 1987). Here parts of the model will be looked at in a little detail as they are basically the same as those used to produce the latest forecasts.

The basic forecasts

Unlike previous forecasts, not only were the 1987 forecasts the result of a more sophisticated approach but they were also presented with a significantly greater level of disaggregation than had been the case in the past. As well as that, some degree of sensitivity analysis was offered. The main forecasts were produced using two different cases. In both cases, the expected real growth of the economy was taken to be 4%. Most of the other exogenous variables such as wage rates and prices were expected to increase at their historical rates. In one case, however, these rates were altered by about 1 or 2% on their historical trends in such a way as to increase domestic timber demand. The main forecasts made by the 1987 LRP are shown in Table 7.2.

The first thing that is apparent from the results is the increasing self-sufficiency ratio that was predicted. The actual figure now stands at

Table 7.2. The demand estimates of the 1987 Long Range Prospects (LRP)

	1987 actual	1994 forecast	1994 actual	2004 forecast
Sawnwood	45	42 - 43	51.0	45
Pulpwood	31	40 - 35	42.4	40 - 35
Panels	15		14.1	19 - 17
Other	4	18 - 17	2.0	5 - 4
Total	94	104 - 99	109.5	108 - 101
Domestic supply	35	45 - 40	24.5	52 - 43
Imports	59	59	85.0	56 - 58
Self-sufficiency ratio	37%	43 - 40%	22.4%	48 - 43%

All figures in millions of cubic metres RWE, figures contain rounding errors.
The first figure in the forecast columns relates to the high domestic production scenario.
source: *Shinrin keikaku kenkyuukai* (1987)

around 20% and so, like the previous forecasts, an increase in domestic supply was predicted but was not forthcoming.

On the demand side, the forecasts show a stabilization in the demand for sawnwood, a slow increase in panel product usage and a modest growth in the demand for pulp and paper. All of these forecasts were relatively uncontroversial and could have been obtained by relatively simple trend extrapolation techniques; although this would, of course, have meant a reduction in the informational content of the forecasting exercise.

It is the domestic supply forecasts, and therefore the import levels, which were the most important factors in these forecasts and remain the most significant factor in the most recent forecasts. It is useful, therefore, to look at these forecasts in a little more detail. The analysis of the other forecasts will be reserved for the discussion on the most recent forecasts when past forecasting performance will cast some light on the most recent forecasting effort.

Domestic timber output

The output from domestic forests was broken down in the 1987 LRP by end-use and by type of forest. The figures are shown in Table 7.3.

Table 7.3. Domestic production in the 1987 Long Range Prospects (LRP)

End-use	Source	1984 actual	1994 forecast	1994 actual
Sawnwood	Man-made	12.6	18.7 - 16.3	14.2
	Natural	6.3	3.8	3.2
Plywood & panels	Man-made	0.1	2.8 - 2.7	0.1
	Natural	1.1	0.6	0.2
Pulpwood	Man-made	1.3	3.7 - 3.4	1.0
	Natural	10.5	9.8	5.2
Other	Man-made	0.5	0.6	0.3
	Natural	2.8	2.9	1.7
Total	Man-made	14.5	25.8 - 23.0	15.6
	Natural	20.7	17.1	10.4
Total		35.2	42.9 - 40.1	24.5

All figures in millions of cubic metres RWE, figures contain rounding errors.
sources: *Shinrin keikaku kenkyuukai* (1987), *Shinrin kihon keikaku kenkyuukai* (1997)

As would be expected the main forecast for the end-use for the timber from man-made forests, which would be almost entirely coniferous, is in sawnwood production. Some was predicted to find its way into plywood and pulp, a trend which is yet to materialize despite the beginnings of the use of conifers in plywood reported in Chapter two. The natural forests were thought to be mainly used for pulp, plywood and fuel. Most of these would be broad-leaved, although the conifers from Hokkaido's natural forests constitute a large part of the figure for lumber coming from natural forests. As the table shows, the supply of timber to both the high and low value ends of market has not been of the magnitude expected.

The forecasting methods for natural and man-made forests differed. The forecasts for natural forests were fairly straightforward, using a simple low felling probability; essentially treating felling of natural forests as a random process. The probability of felling was set at around 2% for age classes less than 15 and 5% for older stands. The more important forecasts for man-made forests, which have the more direct influence on the import predictions, were produced using the so-called gentan probability model which was briefly outlined in Chapter three.

Gentan probability theory

Recall that the basic idea behind the gentan probability model is that there exists a gentan probability function which gives the probability that a forest, planted now, will be felled in a certain age class in the future. Whilst some forests could be modelled by this approach, some would be located in unfavourable situations and so would not be managed in a way amenable to modelling using the gentan probability function. The forests which were still be actively managed form the so-called economic forest group.

The basic forecasting model can be expressed by Equation 7.1 where F_i is the forecast volume from man-made forests in age class i. In the equation, A_i is the total area of forest in the age class, f_i is the felling probability for that age class derived as a conditional probability from the underlying gentan probability function, v_i is the volume per hectare of forests in age class i, and α_i is the proportion of the total forest area in the economic group. Thus, the first term on the right-hand side of the equation shows the forecast of output from those forests which are still being actively managed and are sufficiently close to the markets for them to be economic. The second term shows the output from the uneconomic group and this was modelled using a simple random process where the felling probability, r_i, was taken to be a fairly low percentage. In the equation it is assumed that there are n age classes being considered.

$$F_i = \alpha_i f_i A_i v_i + (1 - \alpha_i) r_i A_i v_i \quad \text{..... for } i = 1 \text{ to } n \qquad 7.1$$

The advantage and strength of the gentan probability model for forecasting can be seen from this equation. First, using the information about forests and age class distributions which the Japanese have, the forecasts can be broken down by location, species and age class. This latter means that the forecasts can be used to generate outputs of different forest products and even log assortments could be derived. In the 1987 LRP the mean and variance of the gentan probability functions for the main species were assumed to remain fairly constant into the forecasting period, except in the case of pine where the mean felling age was expected to increase (*Rinyachou*, 1984). Therefore, the only way in which price can enter the model to form a market-style equilibrating mechanism is through the economic forest group ratio α_i. This is incorporated into the overall model in the form shown in Equation 7.2 where P is an index for the price of saw logs, C is and index of logging and production costs and R is the length of new forests roads.

$$a = -61.7 + 1.3(P - C) + 0.57R \qquad 7.2$$

In this way the price of timber is assumed to operate on the proportion of forests which are in the economic forest group and, therefore, influence overall supply. The overall model for domestic demand prediction is shown in Fig. 7.2 and this is the same as the one used in the most recent LRP.

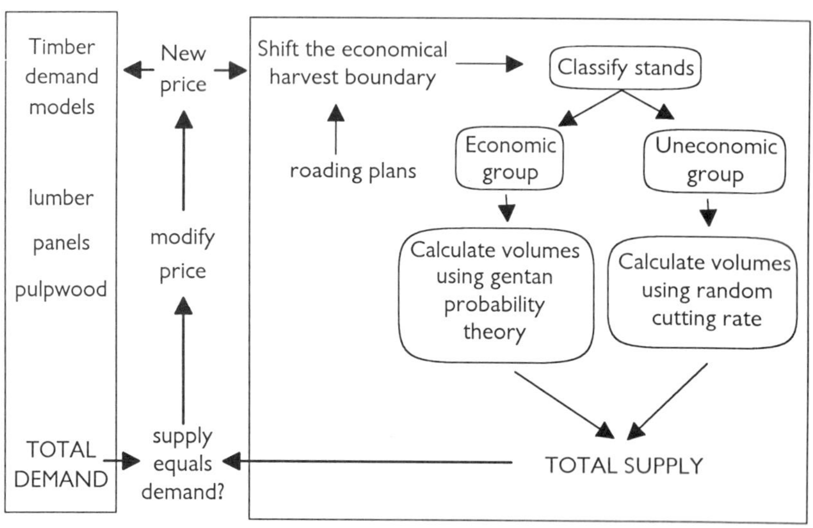

Fig. 7.2. The domestic supply model in the LRP

The 1997 Long Range Prospects

The latest set of forecasts were developed in 1997 using a model very similar to that described above. Once again, two distinct forecasts were produced. One, is called the *iyokuteki* model. The word *iyokuteki* is translated in the dictionary as 'enthusiastic' or 'highly motivated' which are not really words that are usually applied to forecasts. It means here the forecasts which should come into reality if the policies implied by the accompanying discussion are carried out. The word will be loosely rendered as 'optimistic', which if the previous forecasts are an indication, is likely to be a fairly accurate translation. The other forecast produced is referred to as a *suusei* forecast which means 'extrapolative' and here implies that the outputs which will come to pass if the present trends continue. In the English language report on these forecasts by the Japanese (FAO, 1998) the two forecasts are given the names 'potential evolution away from current tendencies' and 'continued current tendencies'.

The main elements of forest policy that have to be fulfilled if the optimistic forecasts are to be accurate are discussed in *Shinrin kihon keikaku kenkyuukai* (1997) and listed in FAO (1998). The main points that relate to the production of forest products are the following. First, concern is expressed about the level of absentee landowners, which is on the increase as Chapter three showed. It is argued that a unification of management is needed by combining the managing of several forests and 'preventing subdivision of forest ownership and management through the acquisition of forest land' (FAO, 1998). This would seem to be an area where the forestry co-operatives have a role. But the restriction on the acquisition of forest land would seem to prevent large private owners, with an interest in forestry, from accumulating forests. Also, it is doubtful to what extent combining forests is a useful exercise. Any economies of scale that can be obtained would require the contiguous blocks of forest actually to be of the same age class. In Japan, in many cases, this does not really occur.

For the factors above to be realized it is necessary to secure the labour force and promote the introduction and development of high-performance forestry machinery which can be effective even in conditions of difficult topography. However, it is difficult to see how this can be achieved. One of the main reasons for the decline of the rural areas in Japan has been the lack of a European-style social security system, with its place being taken by company schemes which have only really been offered by the larger, urban-based firms. It is difficult to see this trend being reversed despite the cost-cutting measures which are almost certain to gain ground in Japanese industry in the future.

Also it is argued that:

> '[a]t the same time, it is important to promote the improvement of living conditions, such as facilities for water supply and household waste disposal. Moreover, while promoting exchange between urban and rural communities through multiple use of forests and smoothness of information distribution, it is necessary to train talent for local leadership' (FAO, 1998).

Exactly what is meant by the last part of the statement is not clear, but the first section presumably relates to the lack of basic infrastructure in a lot of rural Japan, with, for example, the absence of main drainage being the norm rather than the exception. However, the trends in the age of forestry workers shown in Fig. 1.5 in Chapter one show that any policies tackling rural problems have a lot to do. Whilst the LRP do not seem to offer very much by way of concrete proposals for future policy direction, the forecasts produced differ quite markedly depending on whether the policies to counter the problems listed above are actually realized.

The basic forecasts from the LRP

The main output forecasts for the main product groups under the optimistic scenario are given in Table 7.4 and those for the extrapolative case are in Table 7.5.

Table 7.4. The forecasts for the main product groups (optimistic forecasts)

	1992-1994	Lower forecasts		Upper forecasts	
	average	2005	2015	2005	2015
Sawnwood	50.9	49.7	49.8	51.8	53.6
Pulpwood	40.1	44.3	48.4	44.3	48.4
Plywood	14.1	14.1	13.7	15.6	16
Other panels	3.2	3.9	4.7	4.3	4.9
Other	2.3	2.5	2.5	2.6	2.6
Total	110.6	114.5	119.2	118.5	125.5
Domestic	27.4	31.7	36.3	33.6	39.7
Imports	83.3	82.8	82.9	84.9	85.8
Self-sufficiency ratio	24.7%	27.7%	30.5%	28.4%	31.6%

All figures in millions of cubic metres RWE, figures contain rounding errors.
source: *Shinrin kihon keikaku kenkyuuakai* (1998)

The two forecasts contained in the optimistic model differ only slightly in terms of their exogenous variables. In the upper forecasts the price of non-wood building materials was increased compared with the lower forecasts. In the extrapolative model only one forecast was produced. The main assumptions in

Table 7.5. The forecasts for the main product groups (extrapolative forecasts)

	1992-1994 average	2005	2015
Sawnwood	50.9	47.5	44.9
Pulpwood	40.1	45.1	49.5
Plywood	14.1	13.8	13.1
Other Panels	3.2	4.4	5.3
Other	2.3	1.7	1.8
Total	110.6	112.5	114.6
Domestic	27.4	20.0	16.9
Imports	83.3	92.5	97.7
Self-sufficiency ratio	24.7%	17.8%	14.7%

All figures in millions of cubic metres RWE, figures contain rounding errors.
source: *Shinrin kihon keikaku kenkyuukai* (1998)

Table 7.6. The main assumptions in the 1997 Long Range Prospects (LRP)

Item	Optimistic model	Extrapolative model
Non-wood building material price index growth rate	1 or 2% p.a.	1% p.a.
Imported lumber price growth rate	4% p.a.	
Manufacturing wage index growth rate	3% p.a. falling to 2% p.a.	
Housing starts	140 million square metres for first five years 142 million square metres for next five years 145 million square metres for final ten years	
Proportion of wooden housing (area basis)	Maintains present level of 54.1%	Falls to 46.3% over the next twenty years
End-use of timber from man-made forests	Over the next twenty years non-lumber uses will rise to 30%	Non-lumber uses hold their present level of around 10%
Utilization of thinned timber	Rises to 70% over the next twenty years	Maintains present level of around 50%
Forest road construction	4000 km p.a.	2000 km p.a.
Timber production costs	Maintain present relative levels	Rise by 2% a year for the first five years then by 1% a year.
Average felling age	Increases by 5 years over the next twenty years	
Real GDP growth	2.5% until 2000 then 2% p.a.	
Exchange rates	No change on levels ruling at time of forecasts	

source: *Shinrin kihon keikaku kenkyuukai* (1998)

the models are shown in Table 7.6.

The overall level of timber demand in the extrapolative model is lower than that in the optimistic model and the reason for this lies mainly in the area of sawnwood. The demand for pulpwood increases slightly in the extrapolative model, and the demand for other panels increases at the expense of the demand for plywood. But the biggest difference lies in the demand for sawnwood which is over 10% down in 2015 in the extrapolative forecast compared with the upper limit of the optimistic forecasts.

As a result there is a large difference in the self-sufficiency ratio between the two sets of forecasts. In the optimistic forecasts the increase in sawnwood demand is accompanied by an increase in the supply of domestic timber and the self-sufficiency ratio exceeds 30% in the upper forecasts, as close as MAFF gets to predicting an era of self-sufficiency or *kokusanzaijidai*. On the other hand, the extrapolative forecasts predict a continuing decline in the level of domestic production and a concomitant drop in the self-sufficiency ratio.

Table 7.6 reasons for the differences in the forecasts are clear. In the extrapolative forecasts, the proportion of wooden housing falls and so reduces the demand for lumber. The use of thinnings also means that, under the extrapolative model, the expansion of the use of domestic timber into areas other than lumber is cramped. Also, the conditions of felling are much better in the optimistic case, with roading and cost conditions being more favourable. The main external factors in the two models are the same; GDP, exchange rates, housing starts and wage levels and so the differences hinge on factors affecting the forestry sector only, emphasizing that, without some form of government intervention in the shape of roading, etc., the dominance of the Japanese forestry sector by imported timber will continue.

The main equations that make up the model are given in the appendix to this chapter along with a fuller breakdown of the forecast outputs.

Saw log and sawnwood model

The demand for domestic saw logs and sawnwood was modelled separately from that for imported logs and lumber but only one supply function, for domestic timber, was estimated. The demand for domestic saw logs was related to the proportion of wooden housing, the price of substitute building materials, the price of domestic timber and the capacity of domestic processing plants. The demand function for imported timber depended upon the total area of housing starts, the price of imported timber and the capacity of the mills processing imported timber.

The supply function for domestic timber rested on the gentan probability model and the area of economic forest group described earlier. The latter was related to timber price and roading density, as outlined above in Equation 7.2, and the economic forest ratio was thought to fall from 40% in the base year to only 30% in the case of the extrapolative model but rose to 58% by the year 2015 in the optimistic model.

The forecasts are summarized in Fig. 7.3 and the similarity of the 1987 and 1997 forecasts is apparent. The tables in the appendix give greater details of these forecasts but the driver behind the increase in domestic timber supply is coniferous timber originating from man-made forests and so the main area of direct competition here will be North America and, to a lesser extent, Russia. In both the latest LRP forecasts the output of broad-leaved saw logs is predicted to decline. This is largely because the supply of timber from natural forests was forecast using a simple time trend. In the optimistic model the same basic time trend was used but it was argued that the downward path of output would halt at some point in the future and that output would then more-or-less stabilize. In the extrapolative model the trend continued until the end of the forecasting period.

In the optimistic forecasts the self-sufficiency ratio for this sector is forecast to rise to between 46% and 48%. In this case, even though overall demand is predicted to rise, the increase in domestic production is more than sufficient to cause a drop in imports and the predicted import levels are 30.5 million cubic metres in 2005 and 27.5 million in 2015. Therefore, from the point of view of the

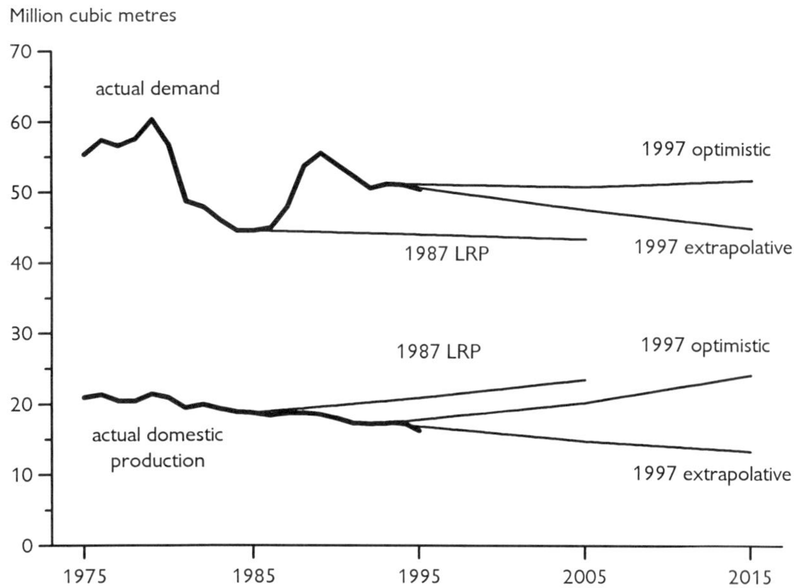

Fig. 7.3. Sawnwood projections in the 1997 LRP

exporting countries, the extrapolative forecasts offer the better scenario. In those forecasts the self-sufficiency ratio is set to fall from its present level of 34% to 29.8%. The declining demand for lumber overall, caused by the increasing substitution for sawnwood in building by non-timber materials and panel products means that, even though domestic production falls, imports also drop from 33.6 million cubic metres to 32.7 million in 2005 and 31.5 million in 2015.

As has already been mentioned, the economic group of forests, these being almost entirely coniferous, was expected to rise to as much as 58% by 2015. This is a level which is similar to that registered in the mid-1970s. The reasons for this are the implied higher levels of roading and the price of timber. Under the optimistic forecasts the price of saw logs is predicted to rise by about 2% by 2005 and 10% by 2015. This change is price is expected to be reflected in the price for sawnwood which is projected to increase by 26% by the year 2015. In the extrapolative forecasts, however, the price of saw logs is expected to fall slightly to begin with and then regain its present level by 2015. However, the biggest difference is to be found in the stumpage prices. In the extrapolative model the stumpage price is expected to continue to decline and to more than halve by the year 2015. However, in the extrapolative model it rises slightly. These are shown in Table 7.7.

If it is accepted that the price level of this type of timber in Japan is going to be determined closely by the level of imports from North America, it is clear that these projections are at the lower end of the range of possible price changes

Table 7.7. The price projections from the 1987 Long Range Prospects (LRP)

	1992 - 1994	Optimistic		Extrapolative	
	average	2005	2015	2005	2015
Stumpage	90.4	90.8	114.0	48.6	36.6
Saw logs	99.0	101.0	109.3	95.6	99.7
Sawnwood	102.7	110.8	127.9	108.4	119.7

source: *Shinrin kihon keikaku kenkyuukai* (1998)

which were discussed in Chapter five. Indeed, the fall in saw log prices in the extrapolative model would need a further strengthening of the yen to be compatible with the rises in the price of North American timber which most forecasts are suggesting.

The forecasts are based on predictions of housing starts which are roughly compatible with those reported in Table 2.2 in Chapter two. There, it will be recalled, a forecast was presented which showed that the number of housing starts was set to fall in the period covered by the LRP. In that chapter it was suggested that the forecasts from the Population Problems Research Centre seemed to err on the side of caution. However, if the number of housing starts in Chapter two are compared with the area of housing starts in Table 7.6 it emerges that, for the two forecasts to be compatible, the LRP is projecting that the average floor area will rise from 96.7 to 108 square metres in 2015. This represents, more-or-less, an extrapolation of the trend for the average floor area shown in Fig. 2.3 in Chapter two. But it would mean that the proportion of the slightly smaller *bunjou* housing would not have to rise dramatically.

Of course, since the time that the model was developed things have changed, notably the collapse of the Asian economies and the fall in the value of the yen. Coupled with this is the effect of the Kobe earthquake on the perceptions of people towards housing. These effects will slow down the demand for lumber because of the recession in Japan but will also make imported timber more expensive. The overall effects of this are difficult to discern. Although the packages announced by the Japanese government at the time of writing are thought to be inadequate to stimulate demand, the tradition has always been to apply expansionary measures in the construction sector. Also, if the price of saw logs increases more-or-less directly in line with the weakening of the yen against the dollar, as seems to have happened in the past (Uusivuori and Buongiorno, 1991), the prices of sawnwood in the future will exceed those predicted in Table 7.7.

The reported forecasting equations do not allow the effects of these changes to be simulated, mainly because the supply side is very complex and cannot easily be laid out; a similar problem to the writer's model reported in the next chapter. Also, the demand functions take housing starts as an exogenous variable, and it is via this variable that economic activity is most likely to influence the demand for sawnwood. The demand for domestic timber is dependent only on the proportion of wooden housing (in terms of housing variables) and it is not clear from the report how this would vary if the economy slows. The demand for imported timber, however, is dependent on the area of housing starts. The implied change in demand for imported timber if housing starts increase by one percent is, *ceteris paribus*, only about 0.22%. On this showing, a slow down in the economy will have only a small effect on the demand for foreign timber. The main factors affecting the level of imported lumber seem to be the price of domestic timber.

Interestingly, the equations show that the demand for imported timber is negatively related to the proportion of wooden housing. As wooden housing, despite the Kobe earthquake, might still be the preferred form, a slow down in the economy could translate into an increased demand for other types of housing and therefore an increase in the demand for imports. A drop of 1 percentage point in the proportion of wooden housing leads, at present levels of demand, to a 0.17% increase in timber imports.

The main conclusion from this section seems to be that the potential for import penetration depends critically on the level of domestic production and, therefore, the behaviour of the Japanese forest owners. This is something which is central to the model developed by the writer to be explained in the next chapter.

Plywood and other panels model

As most of the panels and plywood consumed in Japan are imported, the models describing these sectors are less dependent upon the forecasts of Japanese forest owners' behaviour and, therefore, more straightforward.

Plywood projections

The forecast for plywood demand, in terms of cubic metres, is determined by a simple single demand equation. The main determinants are the area of housing starts, the proportion of wooden housing and prices of plywood and substitutes. The main forecasts are shown in Fig. 7.4.

Most of the raw material is imported but domestic supply becomes significant in the latest LRP. The supply of broad-leaved timber is again thought to decline through time in a similar way to the sawnwood model, indeed the rate of decline is almost the same as that in the lumber model. In the extrapolative model the use of domestic coniferous timber is negligible and that of broad-leaved timber drops to zero by the year 2015. In the optimistic model, the use of broad-leaved timber drops by about half over the forecasting period but the use of timber from the almost exclusively coniferous man-made forests increases to around 1.5 million cubic metres by 2005 and 2.25 million by 2015.

The forecasting equation for plywood does not appear to be wholly satisfactory as it includes a positive price coefficient for plywood logs. This would imply that as plywood logs rise the demand for plywood increases as is

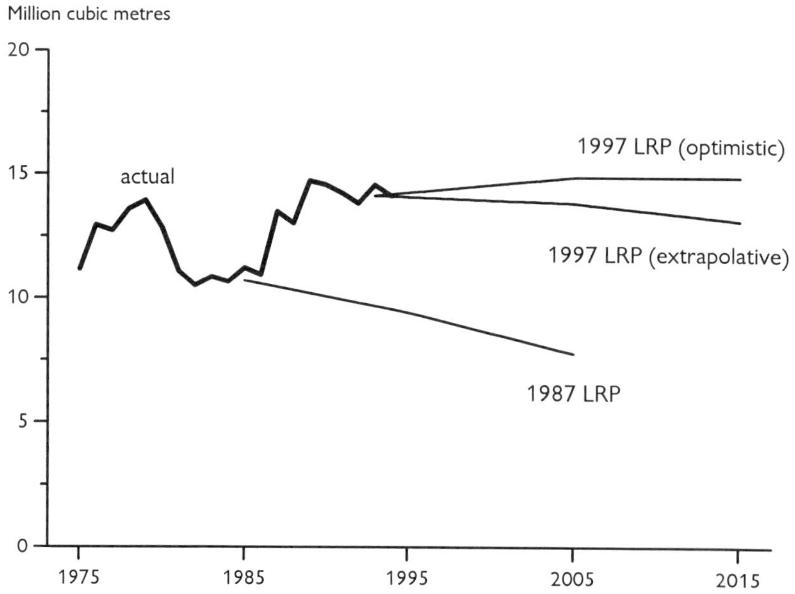

Fig. 7.4. Plywood projections

evident from Table 7.16. No meaningful elasticity measures seem possible from the form of the published equation.

Other panel products

As was shown in the earlier chapters, the demand for other panel products is increasing in Japan. The demand for these products has been projected and the forecasts are summarized in Fig. 7.5. The previous LRP gave forecasts based on a series of data which have subsequently been redefined and are not compatible with the forecasts in the latest LRP and so are not included in Fig. 7.5. However, calculations suggest that the previous forecasts are very similar to those in the present set of projections.

The point to note from the diagram is that the extrapolative forecast projects higher levels of consumption than does the optimistic one. The reason here is mainly a substitution from the preferred plywood into the other panel products which still seem to be deemed inferior to plywood. In total terms, the demand for all panel products rises from 18.95 to 19.65 million cubic metres RWE in the optimistic case and 18.2 and 18.4 respectively in the extrapolative one.

As with the projection for plywood, the demand for domestic timber falls under the extrapolative forecast but rises in the optimistic case with most of the timber being coniferous and coming from man-made forests. The use of factory residues remains constant at half a million cubic metres in the extrapolative case

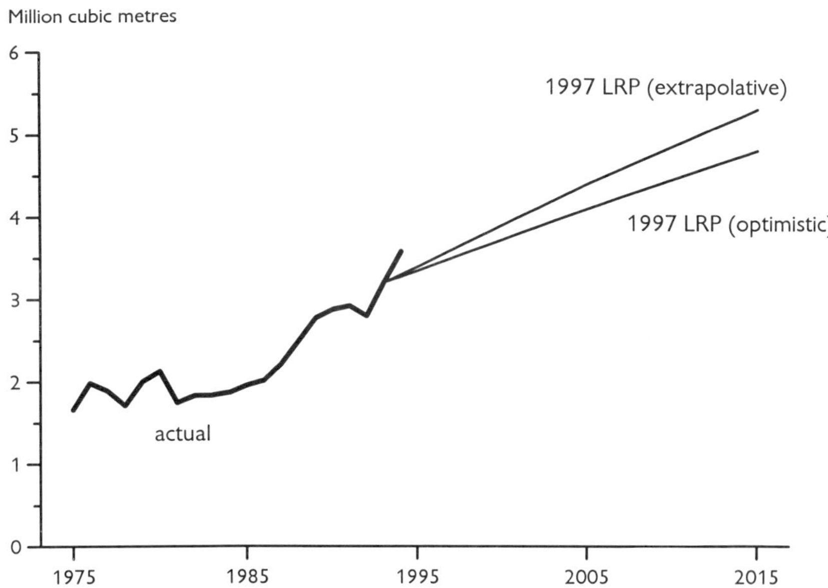

Fig. 7.5. Demand for other panel products

but increases to over one million in the optimistic scenario.

The equations fitted to the data for other panel products imply that demand is dependent upon housing starts, the proportion of wooden housing and the price of substitute panel products. It is also positively related to the price of sawntimber suggesting a substitution between lumber and panels. A very rough calculation suggests that a 1% increase in the price of imported logs would increase the demand for other panels by 0.31%. A similar figure emerges if housing starts increase by 1%.

Pulpwood model

The projections for pulpwood demand are shown in Fig. 7.6. The forecasts for the latest LRP are defined for both the optimistic and extrapolative cases but they are very similar and so only one projection is presented in the diagram. The model for demand is very simple, relating pulpwood demand to the level of real GDP only. The relationship is, according to the equations presented, slowly becoming weaker through time. Factory residues are also important in this relationship. In the extrapolative forecast, 6.2 million cubic metres RWE of factory residues are used and this remains constant over the forecasting period. In the optimistic case this level rises to 7.1 million cubic metres in 2005 and then 8.3 million in 2015.

The use of imported timber rises in the optimistic forecast to 40 million

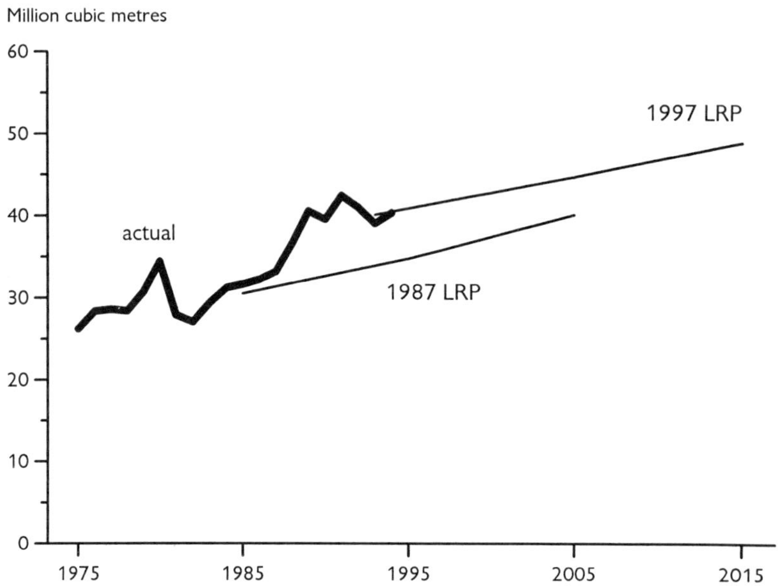

Fig. 7.6. Pulpwood projections

cubic metres RWE in 2015 but in the extrapolative forecasts the level of imports is higher, reaching 47 million cubic metres. The reason for this is the level of domestic pulpwood production contained in the forecasts. In the extrapolative forecast, as was the case with saw log production, the level of domestic production continues to drop in the future. As overall demand is dependent upon GDP the level of demand for pulpwood does not vary greatly between the two forecasts so the drop in domestic production is filled by imports.

The main difference between the two forecasts lies in the level of pulpwood production from man-made forests. In the extrapolative model this falls to 1.1 million cubic metres in the first forecasting period and then stays at this level. In the optimistic model, however, the level of output rises from the current level of 1.4 million to 4.5 million in 2005 and then 7 million by 2015. This is a reflection of the use of thinnings and the non-lumber uses of coniferous timber assumed in the model and shown in Table 7.6. The production from natural forests falls in both cases.

Other timber uses

Finally, the latest LRP deal with other uses of timber. These include *shiitake* bed log production, firewood and charcoal manufacture and what is referred to as 'other industrial uses' which includes the use of poles for scaffolding and power transmission. The present level of demand is 2.3 million cubic metres and this is projected to rise to 2.6 million in the year 2015 in the optimistic forecasts but fall to 1.8 million in the extrapolative ones.

In both forecasts the level of imports are the same, rising from the present half a million cubic metres to 900 thousand cubic metres in 2015. This is mainly in the firewood and charcoal category with no imports forecast for *shiitake* logs.

The origin of imported timber

The LRP present a breakdown of imports beyond the simple classification of 'domestic' and 'foreign' timber suggested above. Although this is not as detailed as it was in the previous LRP it is worth looking at here. The forecasts are based on the following points which are made in the LRP and which are reproduced here (*Shinrin kihon keikaku kenkyuukai*, 1998).

First, it is considered that the pressure of foreign timber on the Japanese market will not diminish into the forecasting period. The reasons given for this are as follows. Although imports show short-term fluctuations there has been a long-term upward trend as shown in the earlier chapters and this can be expected to continue. Also, the number of foreign factories producing timber for the Japanese markets and conforming with Japanese standards is on the increase. The trend to coniferous plywood at the expense of tropical hardwoods, which was discussed in Chapter three, is likely to continue and is certain to cause increased imports of North America, Russian and European timber. It was noted in Chapter three that domestic timber was seen as being somewhat inferior to imported species. The increase in output from Chile and New Zealand, which was discussed in Chapter five, is thought certain to result in increased imports into Japan. Also, the environmental situation in North America is thought likely to lead to an increase in the supply of timber from private forests as the harvest from State lands diminishes. However, it is not noted that the change of species assortment that this might imply would not really suit the present Japanese market. Finally, the establishment of plantations of pulpwood species with high growth rates in the tropics can only cause an increase in imports at the expense of domestic production.

These basic ideas underlie the modelling exercise which was reported above and are roughly consistent with the discussion in the previous chapters and are relatively uncontroversial. In the LRP, figures are given for the optimistic forecasts for various sources of timber. The figures relate only to the optimistic forecasts and represent the average of the upper and lower forecasts contained in that scenario.

North American imports

The figures for North American forest products imports are given in Table 7.8 and the overall picture is one of a slight decline with the most notable falls occurring in log imports and chips. The figures are in line with predictions of

Table 7.8. North American imports in the Long Range Prospects (LRP)

Import	End-use	2005	2015
Logs	Sawnwood	7.3	5.4
	Plywood	0.1	0.1
	Total	7.4	5.5
Sawnwood		14.0	14.6
Veneer		0.4	0.4
Chips	Pulp	7.1	5.6
	Other panels	0.5	0.0
	Total	7.6	5.6
Pulp		10.0	11.5
Other	Other panels	1.1	2.0
	Other	0.1	0.1
	Total	1.2	2.1
Total		40.6	39.7

All figures in millions of cubic metres RWE.
source: *Shinrin kihon keikaku kenkyuukai* (1998)

output from North America and so can be taken to be feasible.

The commentary accompanying the figures in the table merely notes in general that there is likely to be a continued trend in the direction of imports of processed timber products and that the log quality is gradually falling. An increase in the imports of OSB and particle board is expected. It is argued that the volume felled in the USA is below the estimated increment and that this will eventually create pressure to fell more, presumably in the private forest sector. The point is made that the price differential between timber exports from Canada to Japan and to the USA is narrowing, implying that the reduced timber harvests in Canada will tend to be directed more to the USA than Japan.

As noted, the figures in the table above refer to the optimistic forecasts. In the extrapolative forecasts the imports of lumber are predicted to fall at a slower rate and so the extrapolative forecasts imply a greater import level from North America. Similarly, the export of chips and pulpwood would probably increase as the overall projections for imports in this area rise if government policy fails and the extrapolative predictions become the relevant ones. However, the decline in housing starts recently shown in Japan and the environmental pressures in North America have meant that the log imports from North America are already lower than the forecasts for 2005 (Forestry White Paper, 1997).

Tropical timber imports - nanyouzai

The imports of tropical timber in the optimistic forecasts are shown in Table 7.9. In the notes accompanying these forecasts the fact that, despite the desire of most of the countries supplying this form of timber to add value within their own borders, the falling level of growing stock and the increasing demand for

Table 7.9. Tropical timber imports in the Long Range Prospects (LRP)

Import	End-use	2005	2015
Logs	Sawnwood	0.5	0.3
	Plywood	3.5	2.1
	Other panels	0.0	0.0
	Pulp	0.1	0.1
	Total	4.1	2.5
Sawnwood		1.0	0.7
Veneer		7.3	8.0
Chips	Pulp	0.6	0.6
Pulp		0.0	0.0
Other	Other panels	0.2	0.1
	Other	0.4	0.5
	Total	0.6	0.6
Total		13.6	12.4

All figures in millions of cubic metres RWE.
source: *Shinrin kihon keikaku kenkyuukai* (1998)

domestically produced plywood will mean that exports to Japan will not rise rapidly in the future. Although it is thought that the level of plywood production in Indonesia poses no real problems the shift to coniferous plywood in Japan is thought likely to limit any growth in plywoods based on tropical hardwood.

Little comment is made on sources of supply outside of Southeast Asia although it is likely that the authors of the report had in mind the diversification of supply to South America and Africa.

Russia, New Zealand and Chile

The expected imports from these countries are presented in Table 7.10. In the commentary to these figures it is merely noted that the increase in the use of coniferous plywood will call forth an increase in supply from Russia. The use of Russian timber, despite the desirable properties of old-growth larch, spruce and fir noted in Chapter five, does not seem to be sufficient to stop the fall of imports of saw logs from Russia. However, the forecasts are not compatible with the 'modified high growth scenario' in the IIASA forecasts discussed in Chapter six. There, the ability of the Russian East to export timber fell to zero at about the end of the forecasting period here. For the LRP and IIASA forecasts to be compatible there has to be a great deal of investment in the Russian forestry sector or slow internal economic growth.

The fall in demand for timber in the packaging industry is blamed for the drop in saw log demand from New Zealand and Chile. On the other hand, the use of radiata pine for pulp is expected to increase. The possibility is noted that there could be a large expansion of demand in the building area of timber from this source. This would be compatible with the increase in timber harvested from

Table 7.10. Imports from Russia, New Zealand and Chile

Import	End-use	Russia		New Zealand & Chile	
		2005	2015	2005	2015
Logs	Lumber	3.0	1.5	1.2	1.0
	Plywood	1.1	1.0	0.5	0.6
	Other panels	0.0	0.0	0.0	0.0
	Pulp	0.2	0.2	0.0	0.0
	Total	4.3	2.7	1.7	1.6
Lumber		0.9	1.5	0.6	0.5
Veneer		0.1	0.1	0.1	0.1
Chips	Pulp	0.3	0.3	4.9	5.6
	Other panels	0.0	0.1	0.2	0.1
	Total	0.3	0.4	5.1	5.7
Pulp		0.0	0.0	1.7	2.0
Other	Other panels	0.0	0.0	0.1	0.1
Total		5.6	4.7	9.3	10.0

All figures in millions of cubic metres RWE.
source: *Shinrin kihon kekaku kenkyuukai* (1998)

forests with a pruning regime which is foreseen. The projected demand for timber from these sources seems quite feasible given the forecasts discussed in Chapter six.

Other countries

By far the most important import from the 'other countries', which is the final source of timber noted in the LRP, is chips for pulp and pulpwood and this is shown in Table 7.11. The argument appears to be that the diversification of pulp sources will continue with South America and Southeast Asia continuing to be important. It is expected that the imports of chips from Australia will continue but that the species composition will change from natural hardwoods to chips originating from man-made forests. The imports of lumber and saw logs from European sources are not expected to increase greatly because of the competition from the European markets for the products. This last observation seems to be contradicted by the recent rapid growth in imports from European countries, and particularly Scandinavia, which was reported in Chapter six.

Concluding Remarks

The forecasts contained in the latest LRP, and those in the previous one which were based on the same methodology, are a mixture of the straightforward and the complex. The straightforward part is essentially the determination of the level of demand for most of the timber products. The demand for lumber is, uncontroversially, made dependent in the main on housing starts and the

Table 7.11. Imports from other countries

Import	End-use	2005	2015
Logs	Sawnwood	0.1	0.1
	Plywood	0.1	0.0
	Other panels	0.0	0.0
	Pulp	0.2	0.2
	Total	0.4	0.3
Sawnwood		1.9	1.9
Veneer		0.1	0.1
Chips	Pulp	9.1	11.4
	Other panels	0.7	0.8
	Total	9.8	12.2
Pulp		2.1	2.5
Other	Other panels	0.3	0.3
	Other	0.2	0.3
	Total	0.5	0.6
Total		14.8	17.6

All figures in millions of cubic metres RWE.
source: *Shinrin kihon keikaku kenkyuukai* (1998)

proportion of wooden housing, board products are similarly dependent on housing starts and pulp is simply made dependent upon economic activity using GDP. Given the high correlation between GDP and housing starts it is clear that the demand for lumber and panel products is also largely determined by overall economic activity.

The complex part of the modelling comes in the determination of domestic supply. Here the argument appears to be that, although forest owners' behaviour is changing, it is changing only slowly. The mean felling age is increasing only gradually and, therefore by implication, the probability of felling trees in any particular age class is remaining more-or-less constant. The feature that alters the domestic supply is mainly the price of timber, which affects the area of forest that it is economic to fell. This is also dependent upon the level of roading investment undertaken and it is here that government policy has an impact.

The level of total domestic log production forecast by the various LRP is shown in Fig. 7.7. The figures for the 1987 LRP are the average for the upper and lower cases in that forecasting exercise. The conclusion is inescapable that the latest 'optimistic' forecast is likely to be an overestimate. The only forecast that is not overtly optimistic is the present extrapolative one.

The temptation is to argue that the overall levels of demand for timber products might prove to be fairly accurate whilst the forecasts of domestic production are over-estimates, as they have proved to be in the past. Or, more accurately, the level of domestic supply implied by the extrapolative forecasts will be the one which will most likely materialize, whilst the level of demand associated with the optimistic forecasts is the more likely.

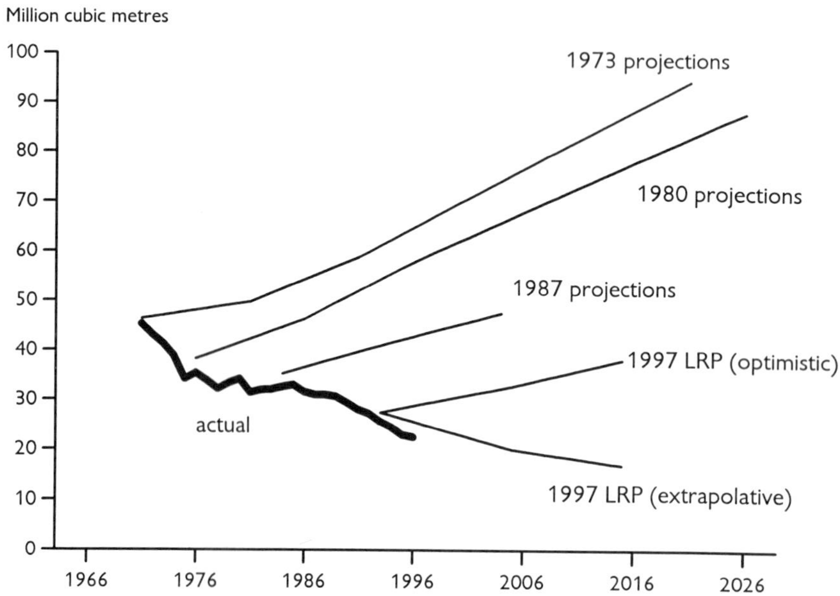

Fig. 7.7. The forecasts of domestic production

If this is the case, the level of imports will differ from those given in the various tables in the previous section. The main effect would be on the level of imports of lumber and saw logs and so the main impact would be felt in North America and, to a lesser extent, Russia.

One of the problems with the forecasting model's structure is the fact that the effect of different growth rates of the Japanese economy cannot be easily tested. Apart from environmental trends within Japan and in the countries supplying Japan, one of the factors which is open to different opinions is the rate of growth of the economy. As this impinges on the level of housing it will have a ripple effect on the demand for lumber, saw logs, panels and plywood.

Also, as will be discussed in the following chapter, the behaviour of Japanese forest owners cannot easily be tested in the model because of the nature of the domestic supply model. This again is likely to be a very important area in the model causing changes in important sections of the forecasts. As has been emphasized in the previous chapters, even a slight change in the behaviour of private forest owners will have a large effect on output. It is to investigate the influence of private forest owner behaviour that the model in the next chapter is presented.

Appendix

The following tables contain the main results of the forecasting exercise discussed in the main body of the chapter. All volume figures are in millions of cubic metres RWE.

Table 7.12. The optimistic forecasts

Demand	Supply	1992 - 1994	2005	2015
Lumber	Domestic	17.3	19.7 - 20.8	22.8 - 25.6
	Man-made forests	13.8	16.8 - 17.9	20.2 - 23.0
	Natural forests	3.5	2.9	2.6
	Imports	33.6	30.0 - 31.0	27.0 - 28.0
	Total	50.9	49.7 - 51.8	49.8 - 53.6
Plywood	Domestic	0.3	1.3 - 1.9	2.1 - 2.6
	Man-made forests	0.0	1.2 - 1.8	2.0 - 2.5
	Natural forests	0.2	0.1	0.1
	Imports	13.9	12.8 - 13.7	11.6 - 13.4
	Total	14.1	14.1 - 15.6	13.7 - 16.0
Other panels	Domestic	0.5	0.9 - 1.1	1.3 - 1.4
	Man-made forests	0.1	0.6 - 0.8	1.0 - 1.1
	Natural forests	0.4	0.3	0.3
	Imports	2.7	3.0 - 3.2	3.4 - 3.5
	Factory residues	0.5	0.7	1.1
	Total	3.2	3.9 - 4.3	4.7 - 4.9
Pulp	Domestic	7.5	8.0	8.4
	Man-made forests	1.4	4.5	7.0
	Natural forests	6.2	3.5	1.4
	Imports	32.6	36.3	40.0
	Factory residues	6.2	7.1	8.3
	Total	40.1	44.3	48.4
Other uses	Domestic	1.8	1.8	1.7
	Man-made forests	0.0	0.1	0.1
	Natural forests	1.7	1.7	1.6
	Imports	0.5	0.7	0.9
	Total	2.3	2.5	2.6
Total	Domestic	27.4	31.7 - 33.6	36.0 - 39.7
	Man-made forests	15.3	23.2 - 25.1	30.3 - 33.7
	Natural forests	12.1	8.5	6.0
	Imports	83.3	82.8 - 84.9	82.9 - 85.8
	Factory residues	6.7	7.8	9.4
	Total	110.6	114.5 - 118.5	119.2 - 125.5

source: *Shinrin kihon keikaku kenkyuukai* (1998)

Table 7.13. The extrapolative forecasts

Demand	Supply	1992-1994	2005	2015
Lumber	Domestic	17.3	14.8	13.4
	Man-made forests	13.8	13.3	12.9
	Natural forests	3.5	1.5	0.5
	Imports	33.6	32.7	31.5
	Total	50.9	47.5	44.9
Plywood	Domestic	0.3	0.1	0.0
	Man-made forests	0.0	0.0	0.0
	Natural forests	0.2	0.1	0.0
	Imports	13.9	13.7	13.1
	Total	14.1	13.8	13.1
Other panels	Domestic	0.5	0.2	0.1
	Man-made forests	0.1	0.0	0.0
	Natural forests	0.4	0.2	0.1
	Imports	2.7	4.2	5.2
	Factory residues	0.5	0.5	0.5
	Total	3.2	4.4	5.3
Pulp	Domestic	7.5	3.9	2.5
	Man-made forests	1.4	1.1	1.1
	Natural forests	6.2	2.8	1.4
	Imports	32.6	41.2	47.0
	Factory residues	6.2	6.2	6.2
	Total	40.1	45.1	49.5
Other uses	Domestic	1.8	1.0	1.9
	Man-made forests	0.0	0.0	0.0
	Natural forests	1.7	1.0	0.9
	Imports	0.5	0.7	0.9
	Total	2.3	1.7	1.8
Total	Domestic	27.4	20.0	16.9
	Man-made forests	15.3	14.4	14.0
	Natural forests	12.1	5.6	2.9
	Imports	83.3	92.5	97.7
	Total	110.6	112.5	114.6

source: *Shinrin kihon keikaku kenkyuukai* (1998)

In the following models the dependent variable is in thousands of cubic metres RWE or, in the case of panel products, thousands of square metres. All indexed variables are formed so that the value for 1990 equals 100.

Table 7.14. Demand for domestic sawnwood

Variable	Units	Coefficient	Significance * 5%; ** 1%
Proportion of wooden housing	Per cent	0.15674	**
Non-timber substitutes, real price index (metals)	Index	0.085338	*
Domestic lumber real price index	Index	−0.0084542	
Manufacturing industry real wage index	Index	−0.0536340	**
Capacity of sawmills specializing in domestic timber	Thousand kilowatts	0.010817	
Dummy	Zero until 1973 One after 1973	−5.2477	**
Coefficient		−0.57288	
R^2	0.98		
Durbin–Watson	1.08		

source: *Shinrin kihon keikaku kenkyuukai* (1998)

Table 7.15. Demand for imported sawnwood

Variable	Units	Coefficient	Significance * 5%; ** 1%
Housing starts	Million m^2	0.035008	
Proportion of wooden housing	Per cent	0.073454	
Non-timber substitutes, real price index (metals)	Index	0.0008851	
Non-timber substitutes, real price index (cement)	Index	0.069522	
Domestic lumber real price index	Index	0.12471	**
Foreign lumber real price index	Index	−0.0000192	
Manufacturing industry real wage index	Index	−0.025566	
Capacity of sawmills specializing in domestic timber	Thousand kilowatts	−0.016131	**
Capacity of sawmills specializing in foreign timber	thousand kilowatts	0.012445	
Dummy	Zero until 1973 One after 1973	−1.8539	
Coefficient		10.991	
R^2	0.98		
Durbin–Watson	2.06		

source: *Shinrin kihon keikaku kenkyuukai* (1998)

Table 7.16. Demand for plywood

Variable	Unit	Coefficient	Significance * 5%; ** 1%
Housing starts	Million m^2	3.0747	
Proportion of wooden housing	Per cent	1.2959	
Non-timber substitutes, real price index (metal)	Index	5.1084	
Non-timber plywood substitutes, real price index (cement)	Index	2.6122	*
Domestic lumber real price index	Index	5.34	
Plywood logs real price index	Index	1.1102	
Manufacturing industry real wage index	Index	–7.4379	**
Real fuel price index	Index	3.1285	
Size of mills using in domestic timber	Thousand kilowatts	–3.1146	**
Raw material use per employee	Thousand m^3	2.1522	**
Dummy	Zero until 1973 One after 1973	–144.89	
Coefficient		1,423.9	
R^2	0.97		
Durbin–Watson	1.70		

source: *Shinrin kihon keikaku kenkyuukai* (1998)

Table 7.17. Demand for non-plywood panel products

Variable	Units	Coefficient	Significance * 5%; ** 1%
Housing starts	Million m^2	0.4749	
Proportion of wooden housing	Per cent	0.8186	
Non-timber substitutes, real price index (metal)	Index	0.2847	
Non-timber substitutes, real price index (cement)	Index	–0.2417	
Domestic lumber real price index	Index	0.5557	
Foreign lumber real price index	Index	0.4761	**
Manufacturing industry real wage index	Index	–0.4102	**
Real fuel price index	Index	0.7988	*
Size of mills using in domestic timber	Index	–0.4065	*
Size of mills using in foreign timber	Thousand kilowatts	–0.1076	*
Raw material use per employee	Thousand m^3	0.3415	*
Dummy	Zero until 1973 One after 1973	–2.3213	
Coefficient		1,423.9	
R^2	0.98		
Durbin–Watson	1.90		

source: *Shinrin kihon keikaku kenkyuukai* (1998)

Chapter eight:

An Alternative Model

Introduction

In the previous chapter the model developed by MAFF was presented. However, despite the work that has gone into the preparation of the most recent forecasts, and those which preceded it, there are problems attached to MAFF's model. Some of these are of a rather technical nature and the arguments will only be summarized here. However, they relate to the behaviour of forest owners and, as has been seen in the previous chapters, this is likely to have a critical effect on domestic production and, therefore, imports. Another problem is the use of housing starts as the main driver for demand in the important sawnwood models. While the arguments in the early part of the book made it clear that housing is the main determinant of lumber demand, it also emphasized that housing demand depends upon the level of activity in the economy. So a better start might be to relate housing starts to, say, GDP and then go on to relate timber demand to housing starts. Whilst this has the weakness of introducing another relationship into the 'front end' of the model and, thus, additional forecasting errors, it has the advantage that differences in macroeconomic conditions can be directly related to the level of timber demand. This will be the approach taken here.

In the following pages a model of the demand and supply for coniferous timber will be presented which allows changes in GDP and the behaviour of forest owners to be more explicitly considered. The objective is not so much to arrive at a set of forecasts but rather to establish a set of limits to the possible outcomes in the market for coniferous timber in Japan. In the earlier chapters it was argued that this market was, in a sense, on a knife edge and the optimistic and extrapolative models in the previous chapter illustrated this. It would seem to be fairly straightforward to produce forecasts which indicate that output in Japan will either continue to decline or increase dramatically. What is needed is a deeper understanding of the conditions that will lead to these outcomes.

Pulpwood demand and the demand for panel products will not be pursued in this chapter. The models fitted by the Japanese are likely to be as good as any in these sectors. Also, somewhat sadly, interesting areas such as the development of the use of domestic coniferous species in the production of plywood are only just beginning to appear in the data and so do not lend themselves to detailed statistical analysis. Thus, whilst the trend away from tropical hardwoods is clearly an important one, it is one which will not be analysed here. The main thrust of this chapter will be the use of coniferous timber in the sawnwood area and so will concern Japanese domestic production and the competition with North America, Russia, New Zealand and Chile.

Forest Owners' Behaviour

The model of forest owners' behaviour underpinning the forecasts of MAFF is based on the gentan probability theory described in Chapters three and seven. In that model, the average felling age and the variance of the felling age distribution describe the felling propensity of forest owners. The economic forest group discussed in the previous chapter divides forests into those being actively managed and those which are, in some sense, too far from the market to warrant continued management. The effect, as was shown in Fig. 7.2, is to create a bifurcation in forests with the division dependent upon price.

The effect is that, if a forest owner has forests that are favourably located, management will proceed in a way that is capable of being described by a gentan probability distribution. On the other hand, if the forests are more distant from the markets, management is modelled as a simple random process. The implication is that behaviour will suddenly change if the price alters. Whilst this aspect of the model might well be realistic, if the forest remains in the economic group the changes in the behaviour of the forest owner can only be modelled by changes in the mean and variance of the gentan probability distribution. Thus, the changes that have been observed in Japanese society, and especially those in the rural economy, which were outlined in Chapters one and three, can only impinge on the models developed by MAFF by changes in the average felling age and its variance.

As Table 7.6 in the previous chapter showed, the present models suggested an increase in the mean and variance of the felling age. The mean felling age was expected to increase by about 5 years over the next 20 years. Although it was not noted in that table, the variance was also expected to increase slightly. The effect of these changes on the felling probabilities is shown in Fig. 8.1. These are calculated using a gamma distribution for the gentan probability function which is the functional form chosen by the Japanese (Suzuki, 1973). The initial probabilities in the diagram are calculated with a mean of ten age classes and a variance of the same, these being similar to the figures used in the 1987 LRP (*Rinyachou*, 1984). Thus, the average felling age is 50 years. The final felling

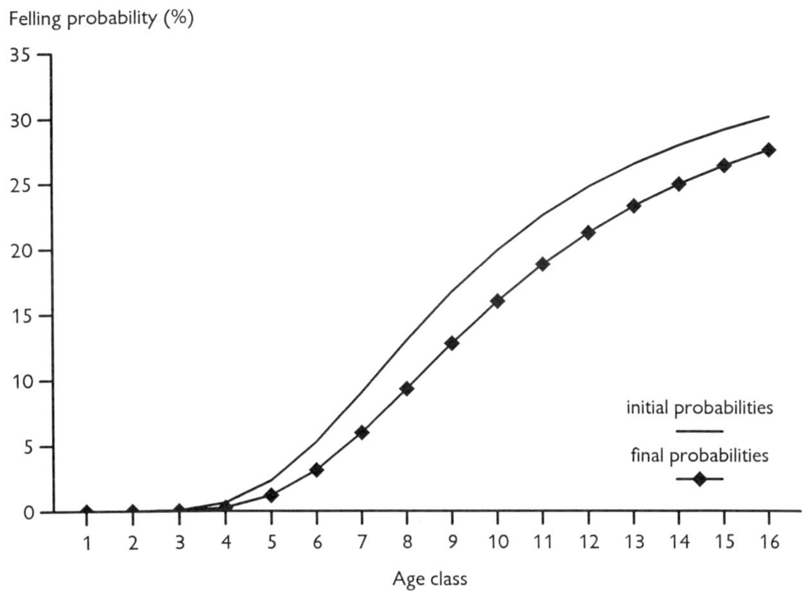

Fig. 8.1. The implied changes in felling probabilities in the LRP

probabilities are those which emerge when the mean increases to 55 years and the variance increases slightly.

The first point to note is that the felling probabilities are considerably greater than those shown in Figs 3.5 and 3.6 in Chapter three. There, the 'probabilities' were empirical ones in the sense that they showed the actual proportions of standing areas which were felled. For example, Fig. 8.1 implies that the felling probability for age class ten in the initial period is around 20%. However, in Fig. 3.6 the felling probability for the same age class for the period 1990 to 1995 was nearer to 5%. The implication would be that the economic forest group ratio is in the region of one-quarter. Therefore, the 20% felling probability in Fig. 8.1 would only be applied to one-quarter of the area of forest and the observed proportion of the area which is felled would be around 5%. The actual area felled will, therefore, be given by the felling probability multiplied by the economic forest group ratio as shown in Equation 7.1 in the previous chapter.

Returning to Fig. 8.1, the changes in the rural economy and the circumstances of forest owners, other than those which can be expressed through the price of timber which moderate the felling probabilities via the economic forest group, are expressed by this shift in felling probabilities. Thus, the LRP seem to be implying that, because of factors such as rural migration, increases in job opportunities in other sectors and the ageing of the rural population, the propensity to fell will decline even if the price of timber does not alter.

It follows that the crucial variables in defining output of Japan's forests are the mean and variance of the gentan probability distribution which will define the felling probabilities. In Chapter three it was suggested that the output from Japan's forests would, if nothing changed, either increase or decrease depending on what was meant by the term nothing changing. If it is argued that the changes in rural society have now worked their way to some sort of stability, the effect would be that the felling probabilities, and hence the mean and variance of the gentan probability distribution, are now stable. On the other hand, if the changes are still continuing the felling probabilities will continue to decrease in a way similar to that shown in Fig. 3.6 in Chapter three or Fig. 8.1 here.

It is as this point, however, that a number of problems emerge. It would seem relatively simple to derive a set of simulations from MAFF's model for different means and variances of the felling ages. This would go some way to answering the question of what effect the changing behaviour of forest owners would have on domestic supply and, therefore, imports of coniferous timber. However, as the writer has shown elsewhere, there appear to be problems in the method employed in the calculation of the mean and variance of the gentan probability distribution (Blandon, 1991b; 1993; 1994). Without going into detail, the mean and variance of the gentan probability distribution need to be calculated using a maximum likelihood approach based on censored sample theory and when this is done it proves impossible to calculate a figure for the economic forest group from the data. Also, it emerges that the mean and variance of the distribution have been moving through time at a rate far different from that suggested in the Japanese forecasts.

Sadly, another problem emerges here. As is all too typical, just at the time when some precision in forecasting is needed, the form in which the data are collected is altered and it proves impossible to establish a consistent time series which includes the latest data. This is what has happened in Japan where the regions which report data for forest areas have been re-defined. Thus, a consistent series of figures for areas felled at the regional level cannot really be constructed after the early 1990s.

However, one data set extending from 1965 to 1995 can be constructed for Aichi Prefecture which is around Nagoya in central Honshu. From these data it is possible, using the methods developed by the writer, to produce a consistent set of estimates for the mean and variance of the gentan probability distribution for sugi and hinoki. These can be related to time and the price of timber and the resulting relationship for the mean of sugi is given in Equation 8.1.

$$mean = -1445 + 0.74time - 0.00028stumpage\ price \qquad 8.1$$

In this equation the mean felling age of sugi is dependent upon time, expressed as the calendar year, and the stumpage price of sugi, in yen per cubic metre. The mean of the distribution, which Equation 8.1 implies, is far greater than that suggested by the Japanese work. The figure for 1990 is just under 28 age classes or 140 years. Because it is impossible to unravel the size of the economic forest group from the data, this high figure presumably represents a form of weighted average of the mean felling ages for those who are managing their forests and those who are not. But, it is considerably greater than the ten age classes or so assumed in MAFF's models. Also, the equation shows that this mean is increasing at a rate of 0.74 years every year. Once again, this figure is presumably an amalgam of the change in the felling age and the change in the proportions of those who are actively managing and those who are not. But it is not consistent with the figures from MAFF which assumed that the mean would change by only one age class over 20 years.

Otherwise, the interpretation of the equation is fairly straightforward. If the price of timber is held constant, the mean felling age increases over the years and this reflects the changes in the rural economy. On the other hand, an increase in the stumpage price of timber means that felling is encouraged as the mean and variance are decreased, so increasing the felling probabilities. Using such an equation it is possible to create a market-style equilibrating mechanism within an overall supply and demand model.

If it is believed that the trends which have been observed in the past in the rural sector will continue unabated the linear time trend implied by Equation 8.1 can be extrapolated into the future. As an alternative model, a similar equation was fitted to the means and variances of sugi but in this case the time variable was treated as the natural logarithm of the calendar year (minus 1945). This function, by its form, means that the effects of time on the parameters of the gentan probability distribution will diminish into the future. This simulates the idea that the increases in the average felling age will not continue unabated into the future but rather will gradually come to an end as the situation in the rural economy reaches some form of long-term equilibrium. If, on the other hand, it is believed that such trends have come to an end now, the time trend can simply be ignored in the determination of future means and variances of the felling age.

A model of output from man-made forests

In line with the discussion above, the following procedure was adopted. As will be seen, it is far from satisfactory. However, given the available data, it seems to be the only way of creating a model of the form implied by gentan probability theory and, therefore, link the behaviour of forest owners with output levels and, by implication, imports. The following explanation will be couched in terms of sugi, but the procedure for hinoki was the same.

First, data for the mean and standard deviation of the gentan probability distribution were derived. This consisted of using standing areas such as those in Table 3.5 in Chapter three and employing maximum likelihood, censored sample methods (Blandon, 1991a,b). The data related to the five different planning areas in the Nagoya region. The surveys used to obtain the data were repeated every 5 years with one planning area being surveyed each year. The equations were then fitted using simple linear regressions and a 5-year average of stumpage prices taken from national data. The equations are shown in Table 8.7 in the appendix at the end of this chapter.

Using the parameters of the gentan probability distributions derived from these equations it was possible retrospectively to 'predict' the area that would have been felled if the gentan probability distribution implied by the fitted equations had occurred throughout Japan. This was achieved by using the age class distribution for sugi for all of Japan as reported in the latest Forestry Census. Yield tables for Aichi Prefecture allowed the area figures to be converted to volumes for 5-year intervals from 1965 to 1995. Annual figures were then calculated on the basis of linear interpolation. Finally, to 'calibrate' the overall forecasts so derived, the actual volumes of sugi felled were regressed against the predicted figures. The resulting regression equation essentially adjusted the 'forecasts' based on Aichi data to fit the national figures. It, therefore, went some way to accommodate the differences in the characteristics of the region from which the data were drawn and the rest of Japan. This equation was then used to define the final set of 'forecasts'. The results of this somewhat convoluted procedure are shown in Fig. 8.2 where the actual and predicted volumes of sugi for the whole of Japan are shown.

Again, data inadequacies meant that for species other than sugi and hinoki the procedure outlined above could not be applied. Time series for the parameters of the gentan probability distribution could not be computed as a long

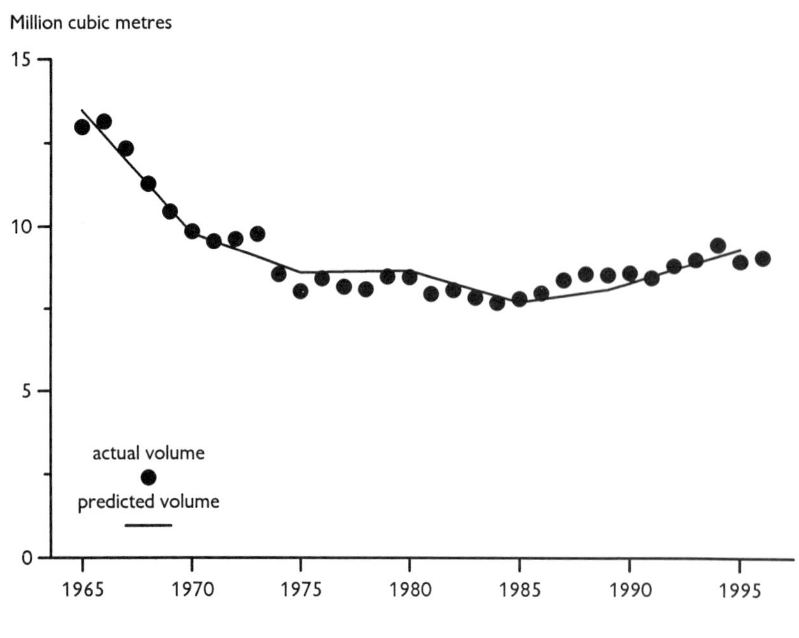

Fig. 8.2. The accuracy of the forecasts of sugi output

series of areas felled was not available. The analysis which could be applied suggested that the mean and variance of the gentan probability distribution for other conifers were moving in similar ways through time to the movements discovered for hinoki and sugi.

The approach outlined above formed the basis of the supply side of the models to be presented. Three models were produced to cover the important coniferous timber markets. The first relates to sugi and *beizai*. Hinoki is treated in a separate model. Other domestic conifers were taken together with Russian, New Zealand and Chilean timber in the third model. Some of the coniferous timber in the latter model derives from natural forests. The underlying idea is that sugi is mainly competing with log and lumber imports from North America, hinoki is seen as being somewhat distinct by consumers and that the other conifers compete against imports of radiata pine and Russian timber.

Sugi and North American Timber

The main model developed relates the levels of economic activity to housing starts and then relates housing starts to the demand for sugi and timber from North America. The underlying idea is that the overall demand for this sort of timber is created by the housing market and that imports of *beizai* essentially fill the gap left between demand and the domestic supply of sugi. The implication is that sugi and *beizai* are seen as perfect substitutes in the housing market. The internal price of sugi is related to the import price of logs from North America. In the forecasts this price depends on the exchange rate and the suggested North American domestic price levels discussed in Chapter three. The overall form of the model is shown in Fig. 8.3 and the various sub-models will now be examined.

Housing models

Models of the demand for housing were developed for three categories, *mochi-ie*, *bunjou* and, thirdly, *kashi-ie* and *kyuu-you* taken together. Attempting to predict the levels of building starts for *kashi-ie* and *kyuu-you* separately did not prove very successful. First, the floor area of housing starts in each of these categories was predicted. Then the proportion of wooden housing in each category was forecast to give the overall area of wooden housing.

Housing starts defined by floor area

The equations concerning floor areas related housing starts to GDP levels, land prices and interest rates. The equations here included the variable interest rates multiplied by the land price index to simulate the cost of housing. The equations which were fitted are shown in Table 8.8 in the appendix. The results suggest that *kyuu-you* and *kashi-ie* housing are seen as being inferior to the other categories

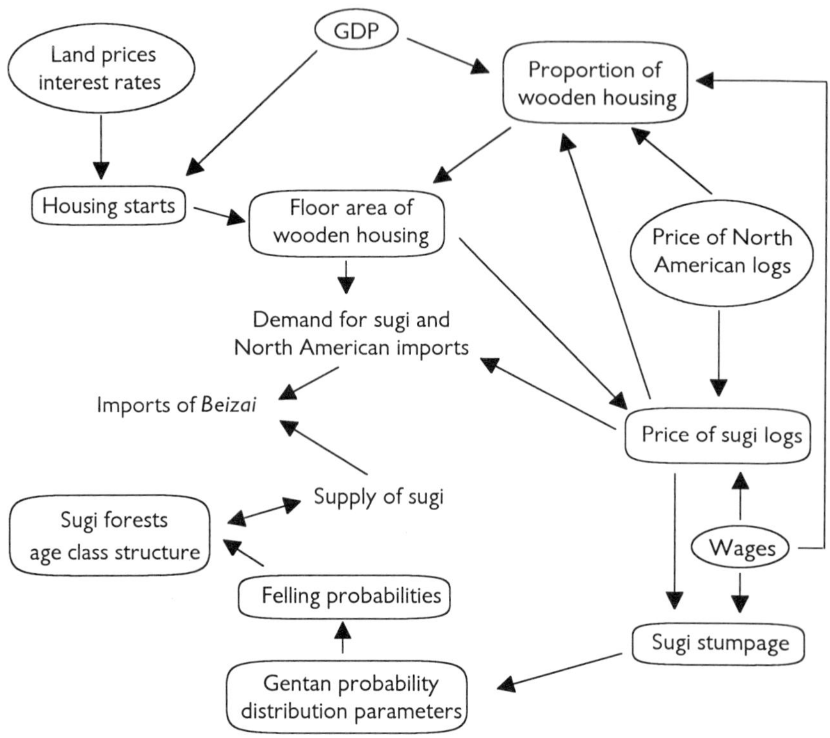

Fig. 8.3. The sugi and *beizai* model

as their area is negatively related to GDP and positively related to interest rates suggesting that a decrease in the GDP or an increase in interest rates drives people from the other preferred forms of housing into rented accommodation.

The main assumptions applied to the exogenous variables in the model are shown in Table 8.1 and all of the figures refer to annual percentage changes. Thus, real interest rates were assumed to rise by 1% each year from 1995 to 2000, not by one percentage point each year. The resulting forecasts on the basis of the exogenous variables shown in Table 8.1 are given in Fig. 8.4. These forecasts seem to be more optimistic than both those in the LRP and those reported in Chapter two. The overall floor area in Fig. 8.4 is 151.5 million square metres in 2000, 160 million in 2010, falling back slightly to 155.9 million in 2020. Whilst the pattern is in line with the predictions of total housing starts in Table 2.2 in Chapter two, the areas are more than 10% greater than the figures assumed in the LRP and shown in Table 7.6. Those stood at between 140 million and 145 million.

In order for the forecasts from the Population Problems Research Centre for the year 2010 shown in Table 2.2 to be compatible with the forecasts in Fig. 8.4

Table 8.1. The main assumptions in the models

	1995 to 2000	2001 to 2010	2011 to 2020
Real GDP	1%	0.5%	no change
Land prices	−1.5%	0.5%	0.5%
Inflation	2.5%	1.5%	1.0%
Real interest rates	1.0%	0.5%	no change

source: adapted from the forecasts by the *Nihon keizai kenkyuu sentaa* as reported in *Nihon keizai shinbun*, 26/2/1998.

the average floor area implied is around 120 square metres. Given the falling level of the relatively small *kyuu-you* and *kashi-ie* housing, this average figure is not unreasonable. Having said that, however, recent events in the Japanese economy mean, that from the short-term viewpoint the forecasts here look like overestimates.

The proportion of wooden housing

The proportion of wooden housing was estimated for each of the categories in Fig. 8.4. It was determined variously by the average price of sugi and *beizai* logs, the wage level, the price of substitute building materials and GDP. The equations fitted are shown in Table 8.9 in the appendix and the main assumptions about the exogenous variables in this part of the model are shown in Table 8.2.

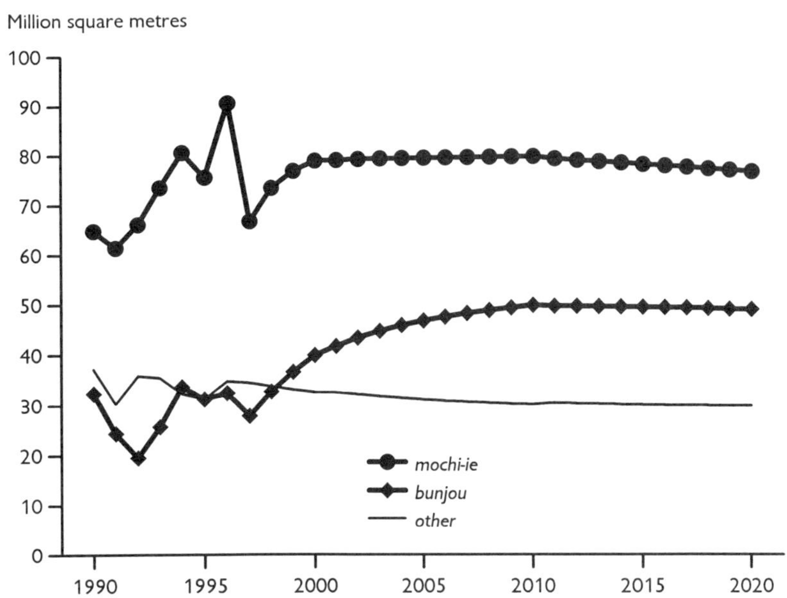

Fig. 8.4. Forecasts of housing starts

The forecast for the overall *mokuzouritsu* or proportion of wooden housing remains roughly constant, falling only slightly over the forecasting period and this is shown in Fig. 8.5 along with the proportion of *bunjou* housing in the overall wooden housing category.

Table 8.2. The main assumptions in the building model

	1995 to 2000	2001 to 2010	2011 to 2020
Wage levels	1%	0.5%	0
Price of substitute building materials	1%	1%	1%
Exchange rate against the dollar	5% devaluation	no change	no change
Beizai log price in North America	3%	2%	0

The results are fairly uninteresting as they predict that relatively stable values for the proportion of wooden housing. The figure for *mochi-ie* falls from its present level of 74% to just under 70% in 2010 and then recovers slightly to 73% in 2020 mainly because of the increasing price of substitutes. The ratio for *bunjou* housing rises from its present level of around 43% to 48% in the early part of the next century, only to fall back to slightly less than its present level by 2020. The proportion of wooden housing in the 'other' category falls steadily from its present level of 18% to around 15%.

The implication seems to be that the long decline in the proportion of wooden housing shown in Fig. 2.5 in Chapter two is coming to a halt. The

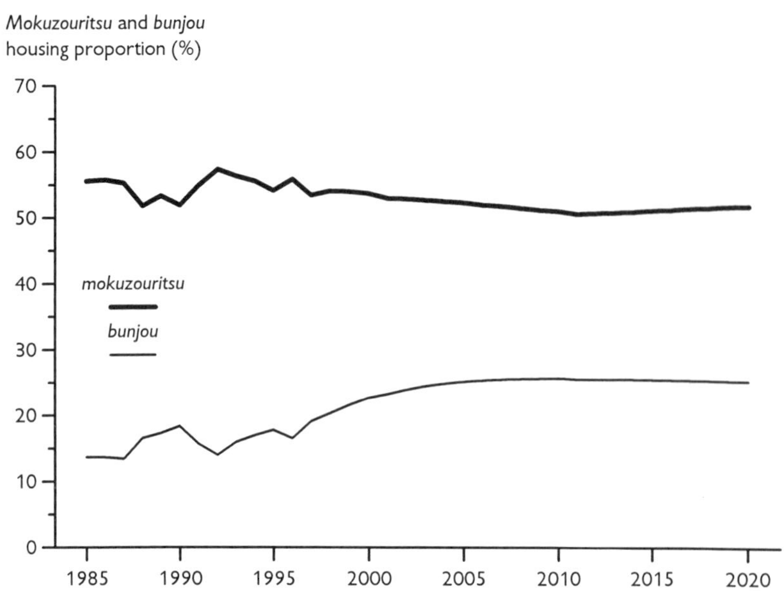

Fig. 8.5. Forecasts of the *mokuzouristu* and proportion of *bunjou* housing

proportion of *bunjou* housing is also of significance as this is an area where lower quality wood and panel products are likely to find it easier to increase their market share. Although two-by-four housing is not synonymous with *bunjou* housing it is likely that the stabilization of the proportion of *bunjou* housing suggests that there are limits to the market penetration of non-traditional building methods and, therefore, to the introduction of panel products.

The overall effect of these forecasts is that the predictions for the floor areas of the different categories of wooden housing look very similar to the patterns shown in Fig. 8.4 for housing starts overall except, of course, that the area is reduced by the proportion of wooden housing. Thus, the size of the market for sawnwood in building would seem to be fairly stable with no real basis for expecting overall growth. Of course, the changes within the market might well mean a change in the type of timber being used but even those changes do not look as if they will be very extensive.

Timber demand

The area of wooden housing is the main driver for the model describing the demand for sugi and *beizai*. The dependent variable in this model is simply the sum of the log production of sugi, the log imports from North America plus the lumber imports from that region converted into RWE using the conversion factor on page *ix*. The explanatory variables are the area of wooden housing starts, the average log price and the price of substitutes materials. Thus, a number of the variables were shared with the model describing the proportion of wooden housing. The equation is given in Table 8.11. Of course, estimating a single equation to model demand is problematic but, given the complexity of the supply side of the model, the greater precision of more sophisticated methods is foregone in order to create a supply model based on the felling probability concept.

The overall argument is that the decision to build a house is made on the basis of income, as measured mainly by GDP. Then the decision is made about the type of house, wooden or non-wooden, on the basis of timber price and then, once the decision has been made, the amount of wood in the wooden house is dependent upon the price of timber.

The results for the demand for sugi and *beizai* using the values for the exogenous variables shown in the previous tables are presented in Fig. 8.6 as the base case. As a sensitivity analysis two other scenarios were tested, a high growth and low growth one. The assumed annual growth rates of GDP for these cases are shown in Table 8.3.

Increasing the assumed rate of economic growth has the effect of increasing overall demand. This it does by increasing the demand for *mochi-ie* and *bunjou* housing but reducing the demand for the other category of housing. As this other category is predominantly non-wooden in make up it follows that a reduction in this area does not reduce the demand for timber greatly whereas an increase in

Table 8.3. Growth rate assumptions in the different scenarios

	1995 to 2000	2001 to 2010	2011 to 2020
Low growth	−1%	−0.5%	0
Base case	1%	0.5%	0
High growth	2%	1.5%	1%

the predominantly wooden *mochi-ie* sector will increase the demand for sawnwood.

Supply of timber

The basic assumptions in the models is that *beizai* and sugi are perfect substitutes. This implies that the prices of their sawnwood products such as pillars ought to be the same. In reality, of course, their prices differ and recently the price of sugi pillars has been falling, as has been reported in Chapter three. However, the log prices do not need be the same as differences in the efficiency of sawmills dealing with domestic and foreign timber will lead to differences emerging in the prices of the raw materials.

Here the evolution of the price of *beizai* logs was taken from the discussion in Chapter five and their prices inside Japan were calculated using the assumed changes in the exchange rates shown in Table 8.2. The log price of sugi was deemed to depend upon the price of imported logs and the relationship

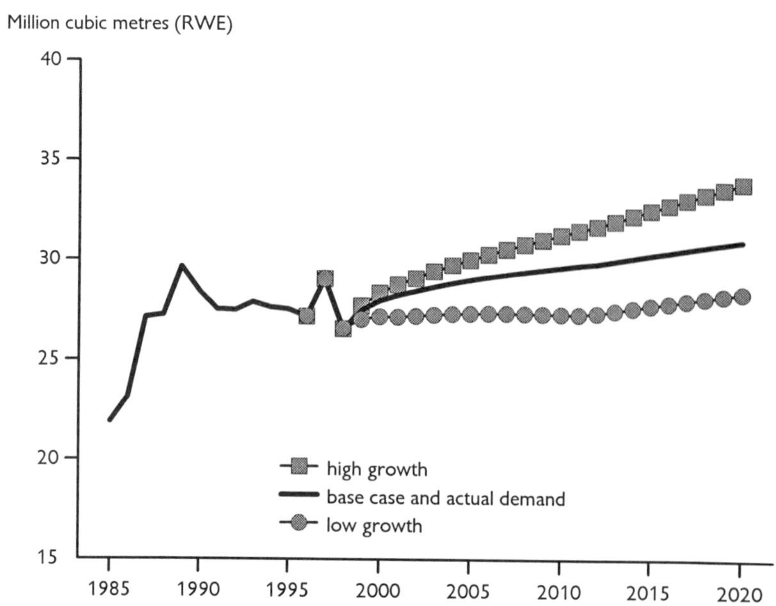

Fig. 8.6. The demand for sawnwood

established is shown in Table 8.10 the appendix. It derives from the assertion in Chapter three that the use of sugi is now not really dependent on its 'Japanese-ness' but rather on the fact that its price has dropped allowing it to compete with the cheaper American imports. Thus, the relationship derived shows that the relative price of sugi logs compared with North American log imports depends upon the proportion of the total housing starts accounted for by *bunjou*. The implication is that as the visual characteristics of timber become less important the price of sugi logs will tend towards those of *beizai* imports.

The gentan probability model which drives the supply of sugi requires the stumpage price as an input. This is derived from the log price using the equation reported in the appendix. The stumpage price is related to the log price and negatively to the wage level, the argument being the obvious one that high wages imply high felling and extraction costs and so drive a wedge between the log price and the stumpage price which the forest owner can obtain. This is in line with Fig. 3.7 in Chapter three. Thus, the assumed *beizai* imported log price, along with proportion of *bunjou* housing, determines the sugi log price and this, with wage rates, determines the stumpage price which allows sugi supply to be determined.

The import of *beizai* and production of sugi under the base case assumptions and using the model with a linear time trend are shown in Fig. 8.7. The linear time trend model is the most pessimistic in terms of domestic production in the

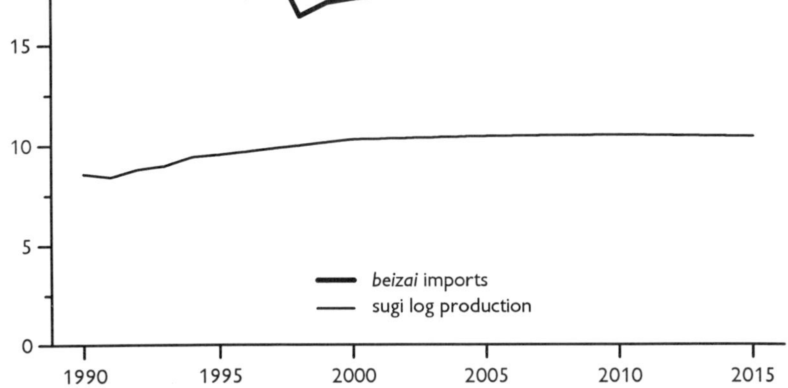

Fig. 8.7. Sugi production and *beizai* imports in the base case

sense that it assumes that the worsening felling behaviour of forest owners will continue at the same rate into the future. Thus, in this model the idea of nothing changing means that the decrease in the management intensity of Japan's forests which has been witnessed in the past will continue at the same rate. The effect is that the output of sugi increases slightly but eventually begins to decline. The output levels in 2005 is 10,491 thousand cubic metres, it peaks in 2010 at 10,541 thousand and then falls back again in 2015 to 10,472 thousand cubic metres. The effect is that imports rise slightly to fill the gap between the slowly rising demand and the stagnant domestic output. The result is, therefore, that the maturing growing stock is not maturing rapidly enough to 'catch up' with the ever increasing average felling age and Japan never enters an era of self-sufficiency. The self-sufficiency ratio for the sugi and *beizai* market remains more-or-less unaltered during the forecasting period.

To simulate the effect of different assumptions on this market the following set of forecasts were produced. The high and low GDP growth rate assumptions from Table 8.3 and high and low log price changes for North American timber were used, the latter are shown in Table 8.4.

Table 8.4. *Beizai* log price growth rate assumptions

	1995 to 2000	2001 to 2010	2011 to 2020
Low growth	0	0	–1%
High growth	5%	3%	2%

The high and low log price growth rates simulate possible extremes to the figures suggested by the discussion in Chapter five. In the low growth case, there is no initial change in domestic *beizai* log prices and an eventual decline as the maturing timber from the southern States reaches the market. In the high growth case, the timber price rises rapidly at first and then the growth rate falls off as extra timber enters the market. In both cases, the price of *beizai* in Japan is found after taking into account the changes in the exchange rates shown in Table 8.2. The results for the years 2050 and 2015 are presented in Table 8.5.

Table 8.5. The results of different assumptions

GDP	*Beizai*	Sugi output		*Beizai* imports		Self-sufficiency ratio (%)	
growth	log price	2005	2015	2005	2015	2005	2015
High	High	10,632	10,746	17,863	18,303	37	37
	Low	9,625	9,676	21,305	24,079	31	29
Low	High	10,905	10,989	14,863	13,878	42	44
	Low	9,916	9,943	18,111	18,959	35	34
High	High	12,996	14,355	15,529	14,694	46	49
	Low	10,495	10,935	20,436	22,819	34	32
Low	High	13,962	15,541	11,806	9,326	54	62
	Low	10,943	11,458	17,084	17,445	39	40

All volume figures are in thousands of cubic metres (RWE).

The figures in the upper half of the table are derived on the assumption that the mean and variance of the gentan probability distribution can be modelled using a linear time trend. The lower half of the table refers to the model using a logarithmic time variable.

The effect of the high GDP growth rate scenarios is to reduce slightly the output of sugi. This may seem somewhat paradoxical as the overall demand for timber increases with increasing GDP. However, the reason for the decline in sugi production lies in the nature of the housing market. An increase in incomes means that housing starts increase in the *mochi-ie* and *bunjou* categories but decline in the rental sector. The overall effect, however, is to increase the proportion of *bunjou* housing. This reduces the requirements for timber with good visual characteristics and so means that, if sugi is to compete, its price must fall relative to North American imports. The effect is to bring sugi prices into line with those of *beizai* and this has a feedback effect on sugi stumpage, thereby decreasing supply. The effects of changing exchange rates and wage levels are similar and are not reported here.

The results are as expected but give a range of self-sufficiency ratios which might be thought of as maximum and minimum levels. The output of sugi is worse under the linear model for generating the parameters of the gentan probability distribution and imports are, therefore, highest under the linear model. Under the logarithmic model, shown in the lower part of Table 8.5, the decline in management intensity is moderated by the assumption that the speed with which the average felling age changes is falling through time. The result is that the maturing growing stock 'catches up' with the felling age and is felled. Here, of course, the levels of imports are at their lowest.

The best situation for the Japanese industry is if GDP grows slowly but the price of logs in North America increases. This means that the market for *bunjou* housing is relatively disadvantaged and sugi can hold its own in the market as a 'Japanese' timber because the relative level of *mochi-ie* housing is greatest. The price of *beizai* logs increases and this has an effect on sugi stumpage, calling forth a greater supply. On the other hand, the worst scenario from the Japanese point of view is if GDP grows rapidly causing a relative expansion of *bunjou* housing and the price of logs in North America does not increase. In this case, the market changes to accepting relatively lower quality sawnwood products for the concealed timber in western-style housing and the import prices do not rise, meaning that the Japanese producers are doubly disadvantaged.

The upper and lower limits of production and imports for the various cases in Table 8.5 are shown in Fig. 8.8. The left-hand part of the diagram shows the situation of high economic growth and low log prices. The right-hand side shows the best situation for Japanese industry of relatively low economic growth but high log prices.

A comparison of the forecasts in Table 8.5 and the forecasts of the LRP produced by MAFF shows that, as far as sugi is concerned, their extrapolative forecasts showing a continuing decline are difficult to reproduce. Even in the

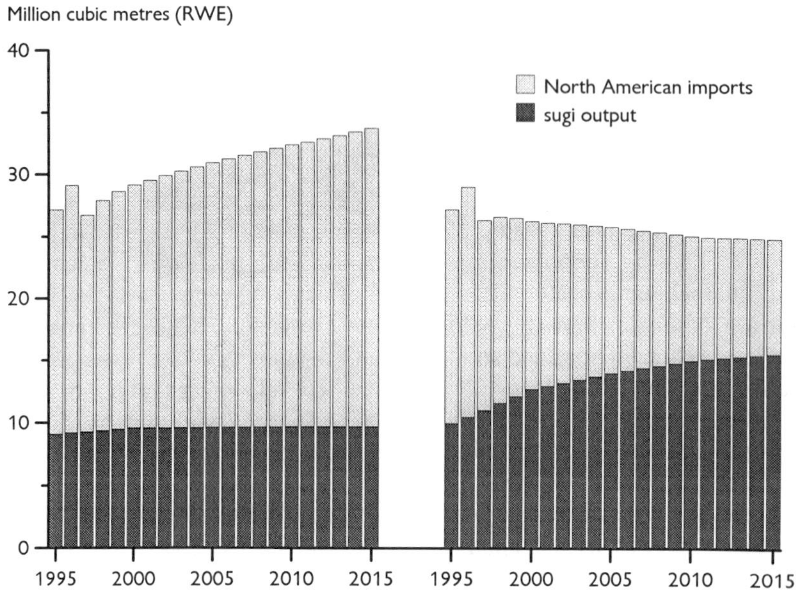

Fig. 8.8. Limiting forecasts of sugi output and imports

high GPD/low *beizai* growth rate scenario output increased slightly. Thus, a decline in sugi production would seem to require the assumption that the situation in the rural communities is going to get worse at a rate which is faster than that seen in the past. Otherwise, it seems very difficult to predict continually falling sugi output. The age class structure of the growing stock already tends to dominate the falling propensity to fell which the linear model of the felling age implies. If the coefficient of the time trend of the mean of the gentan probability function shown in Table 8.7 in the appendix is increased, implying that the mean felling age is changing more rapidly through time than the data from Aichi Prefecture implies, forecasts such as the downward trend of output shown in Fig. 7.7 can be reproduced. But if this adjusted coefficient is used 'backwards' to predict the level of output in the years from 1965, the fit is considerably worse than that shown in Fig. 8.2. Thus, forecasts of declining output are not easily simulated using a function for the time trend of the mean felling age which fits past data. So the extrapolative model of MAFF seems to imply that the deterioration of forest management in Japan will not just proceed at the same rate but must actually accelerate. Indeed, it seems to require a virtual collapse of level of management in the vast bulk of Japan's forests.

An additional point, which makes it difficult to predict falling output, is that any choice of parameters that choke off felling in the early years merely serves to increase the areas of standing timber in the older age classes. Thus, the output which is kept from the market in the early forecasts emerges at a later time rather than not at all.

To take an extreme view, a simulation was undertaken which was based on the logarithmic model but made the assumption that the felling probabilities did not alter from their 1995 levels. This is one interpretation of the idea of nothing changing into the future. The output of sugi in 2005 was only slightly higher than the logarithmic case in Table 8.5 as, over the relatively short period from 1995 to 2005, the increase in the mean of the felling age, which the logarithmic function implied, would not have been very marked. However, output of sugi in 2015 was as high as 18.32 million cubic metres. If this scenario is combined with the low GDP growth rate and the high growth rate of North American log prices, the result is that imports fall to only 6.5 million cubic metres and the self-sufficiency ratio rises to just under 74%.

The result of this section is rather an important one. It is possible, by altering the assumed behaviour of Japan's forest owners via the parameters of the gentan probability distribution, to argue that the output of Japan's sugi forests will stagnate and that imports will continue to dominate, or that Japanese output will increase and imports will be forced down to less than half total market demand. Both of these extreme scenarios, and the other ones which lie in between, can be argued with a fair degree of conviction.

So we are left with the unsatisfactory conclusion that we had at the outset. The market for sugi is on a knife edge and could 'go either way'. It is here that the decision of the Japanese to redefine the planning areas for their forests is especially regrettable. The recent changes in the felling ages are needed to be able to decide which of the various forecasts in Table 8.5 can be viewed with greatest confidence. Rather disturbingly, rough calculations on recent data from Aichi Prefecture suggest that the mean of the felling age has dropped quite radically and if these unadjusted data are input into the model above very large outputs are generated.

Logs and lumber imports

The figures for imports referred to above are all in roundwood equivalents. One of the features of imports from North America which was noted in Chapter five, was the increased imports of sawn timber at the expense of logs. To model this a simple relationship between the proportion of sawn timber in the total import mix and exchange rates was formed. Also, as an explanatory variable, the proportion of two-by-four housing appears (see Table 8.12 in the appendix) suggesting that the increase in modern building methods supports the import of processed timber instead of logs. The forecasts for the proportion of sawnwood, expressed as the ratio of sawn timber in RWE to total imports is shown in Fig. 8.9.

The assumption in Fig. 8.9 is that two-by-four housing will increase its proportion by 10% a year for the first 5 years, then by 5% a year until 2010, and then remain a stable proportion of the housing market. As the diagram shows, it is expected that by around 2010 the log imports to Japan will have fallen to a

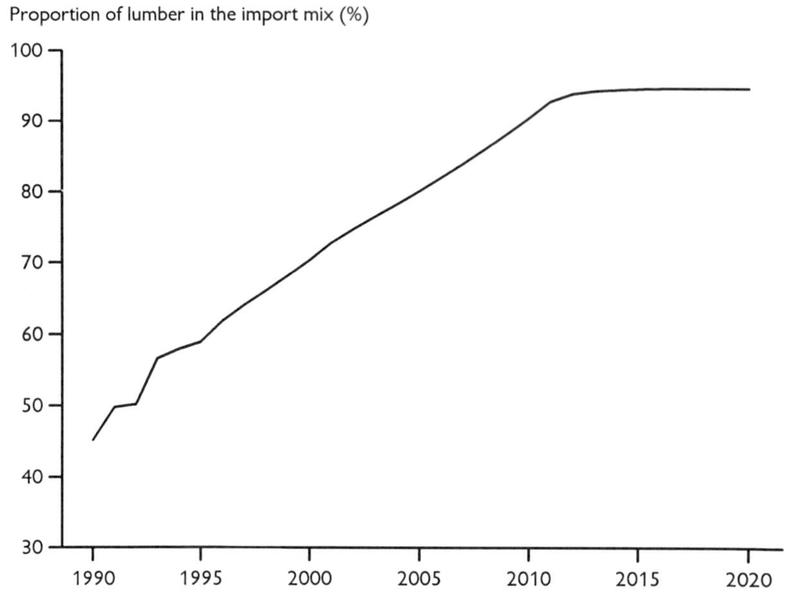

Fig. 8.9. The proportion of lumber in North American imports

very small level. Thus, the imports, being almost entirely sawnwood, will be less in volume terms than those in Table 8.5 where the figures were in RWE.

Although the discussion here has been about imports from North America it is likely that the analysis applies to the imports of sawnwood from Scandinavia which were mentioned in Chapter six. The effect would be that the import figures in Table 8.5 would represent a market in which the USA, Canada and Europe compete. The extra European element would serve to decrease the presence of *beizai* in the market and so the projected import figures given above can, from an American point of view, be seen as likely maximums under the various assumptions.

Hinoki Demand

The modelling effort here treated hinoki somewhat differently from sugi. Because of the price differential and usage of the species, hinoki was considered separately from imports. The supply side was similar to sugi in that a set of equations for the mean and variance of the felling age were fitted using data from Aichi Prefecture in the same way as described above for sugi. The resulting equations are to be found in Table 8.7 in the appendix. In this case only the linear form of time trend was used in the estimations.

The demand side was treated as a single equation with the independent variables being the ratio of hinoki and sugi log prices, the level of wooden housing starts and the ratio of two-by-four housing. It was found that the demand for hinoki was negatively related to its relative price, positively to the area of wooden housing starts and negatively to the ratio of two-by-four housing. The forecasts were then produced using the predictions of sugi log prices which derive from the base scenario described in the tables above. The log price of hinoki was related to its stumpage price in the same way as for sugi, with the relationship depending on the wage rate. Using an iterative method it was possible to find prices which equated the supply of hinoki to its demand and these, along with the corresponding the equivalent log volumes, are shown in Fig. 8.10.

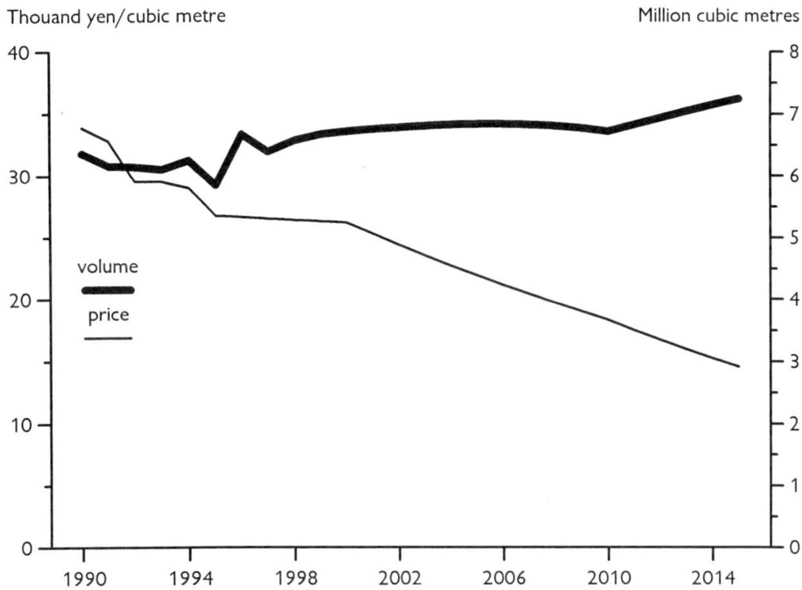

Fig. 8.10. The market for hinoki

The result of the exercise is a relatively stable demand for hinoki but a plunging nominal log price. The demand for hinoki, in its present guise as a quality product at the top end of the market, is limited by the housing market where the trends are moving away from the use of such timber. The increase in the amount of wooden housing helps to create a demand but the assumed increase in two-by-four building methods serves to limit any expansion. However, the supply situation is such that, unless the price drops radically, an enormous oversupply occurs. The fall in price goes some way to increasing demand but mainly helps to choke off supply.

The implication of this forecast is that the present situation in Japan is going to change. The reason is that the price implied by Fig. 8.10 will, fairly soon, fall below the price of sugi and imported logs. The nominal price of sugi, under the same assumptions which gave rise to the diagram here, begins to exceed that of hinoki in the year 2013. Already such a trend might already have begun as was noted in Chapter three. But, if the forecasts here are correct, it implies that hinoki will not be able to maintain its premium price and this is likely to alter the perception of the timber in the market. If it becomes perceived of as 'just another building timber' the output will begin to compete more directly with sugi and *beizai* and so should, more properly, be treated in the previous model. The result might well be to increase the competition on North American timber and so reduce imports below the levels suggested by Table 8.5.

Neither is this situation avoided if the exogenous variables are chosen to favour the market. If high economic growth is chosen the level of *mochi-ie* in the housing market rises and the demand for hinoki increases. However, the effect is to support the price only slightly in the early years of the forecasting period. Indeed, the overall pattern for the market for hinoki remains similar to that shown in Fig. 8.10 regardless of the assumptions made, reflecting a relatively shallow market with inelastic demand and an increasing supply potential which requires a drop in price to stave off overfelling. The pattern, then, is that the premium position of hinoki will not be maintained in the future if the demand and supply patterns which have ruled in the market until now continue unchanged.

Also an 'extreme' forecast could be produced by assuming that the felling probabilities do not change from their present values. In this case, however, the supply of hinoki increases compared with that shown in Fig. 8.10 and so the price is forced down even more and even sooner. Thus, the outcomes in Fig. 8.10 represent a 'best case' for the Japanese timber and the overall conclusion seems that, sooner or later, the price of hinoki will fall.

The production of a forecast in the spirit of the extrapolative forecasts would need a drop in the felling probabilities in excess of those which are implied by the linear trend in the mean of the felling probabilities through time. So, like the sugi market, a simple extrapolation of trends hides the implicit assumption that the decline in management intensity will accelerate rather than stay the same. Thus, the extrapolative forecasts of MAFF discussed in the previous chapter, far from implying that nothing changes, in fact seem to imply a large change in the felling behaviour of forest owners.

Even under the relatively favourable assumptions for this market that the decline in management intensity continues at the present rate, the drop in the price of hinoki is difficult to justify theoretically. Over the 20 years of the forecasts the price drops, in nominal terms, at a rate of about 3% a year and in the final 15 years this rate increases to about 3.8%. This implies a certain irrationality on the part of forest owners as, faced with such a forecast, it would become rational to fell the forests sooner to get a higher price for the product. Thus, those owners with forests in the relatively older age classes, the growth rates of which would be below the implied rate of price drop, should fell immediately. This would serve to depress the price immediately. Viewed from this more theoretical perspective it follows that the market for hinoki is likely to change compared with the situation seen in the years since the Second World War.

Other Conifers

The data for the other conifers are much poorer than for hinoki and sugi. Many planning districts report areas only for the general classification of *matsu* instead of detailed area figures for the various conifers. Often the area of various species is fairly small and the data are subject to large inaccuracies. As a result it is impossible to derive equations such as Equation 8.1 describing the mean and variance of the gentan probability function which underlies the felling probability. Similarly, the stumpage price for the various conifers is not reported and, again only figures for *matsu* are usually to be found. The stumpage price is, however, only about half that of sugi and, as the economics of felling the trees is likely to be very similar, the implied felling activity will be that much less. However, the age class structure of the other conifers is very similar to that of sugi and hinoki and so an increased output could, *ceteris paribus*, be predicted.

As a result of these problems the following approach was adopted. Equations describing the evolution of the mean and variance of the gentan probability function similar to those of sugi were used and the predicted output of 'other conifers', i.e. all coniferous log output except sugi and hinoki, over the period 1965 to 1995 was derived. The time trends of the equations describing the parameters of the gentan probability distribution were optimized to give the minimum squared errors in the predictions. It was assumed that the price variable would affect felling in the same way as it did for sugi. The actual output of logs in the 'other conifers' category and the predicted output are shown in Fig. 8.11.

A single equation for the demand for these conifers and Russian, New Zealand and Chilean imports was then estimated and a similar model to that in Fig. 8.3 was constructed. The demand was found to be negatively related to price and positively related to housing starts in the 'other' category, that is the rented accommodation and the housing provided by companies. It was also found to be negatively related to GDP, the opposite of what has been suggested earlier where it was argued that the use of a lot of the timber for packaging and general uses would mean that its demand would relate to the overall level of economic activity. It was also found to be related to the exchange rate, presumably implying that as the yen strengthens demand rises as imports become relatively more attractive. The equation is reported in Table 8.14 in the appendix.

The forecasts using the base case for the exogenous variables are shown in Fig. 8.11. The decline in this market is apparent and this continues into the future. The domestic supply, however, whilst it declines, does not decline at the same rate and the difference between the two lines in the diagram, which shows

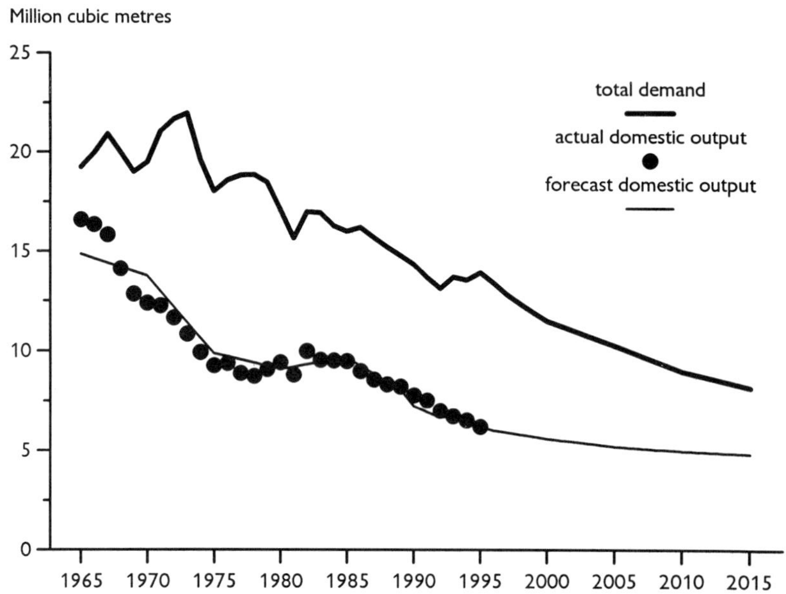

Fig. 8.11. The demand and supply of other conifers

the imported timber levels, diminishes into the future.

The background to this forecast is that the real log price declines slightly into the future. The growing wage rates, modelling the costs of felling and transporting, continue to increase and so the felling intensity for these species diminishes at a faster rate than it does for sugi or hinoki. The result is that, even though there is a maturing stock of timber, the more rapidly increasing mean felling age implies that this is not translated into increases in actual supply. Total supply merely levels off. The import level is, however, squeezed somewhat because of the decreasing overall demand. Thus, somewhat paradoxically, the self-sufficiency ratio increases for this set of species, from about 45% now to just under 60% in 2015.

Given the decline in demand forecast for this market it is difficult to produce forecasts which predict increasing foreign supplies into this area. The maturing timber stock means that only very extreme assumptions about the reduction in management intensity choke off potential supply at a rate which is significant enough to generate increasing imports. Those assumptions which are capable of doing so mean that the predictions of the model for the years from 1965 to 1995 are very inaccurate. Thus, as was the case with sugi and hinoki, an extrapolative forecast suggesting a continuing decline in output requires the assumption of an accelerating decline in management intensity, although not at the rate required for sugi and hinoki.

An extreme forecast, such as that produced for sugi and hinoki, which assumes that the felling probabilities have now stabilized and remain at their present levels, was produced. Here the output of other conifers becomes so large that by 2013 there are no imports in the market at all. In this case, if nothing changes in the market the implication is that the self-sufficiency ratio rises to 100%.

The overall impression is one of declining demand levels and this suggests that countries such as Russia, New Zealand and Chile will have to employ some form of extra, aggressive marketing to expand the market or move to panel products if they are going to find an outlet for their timber. If past experience is a guide, the ability of the Russians to market their products is considerably lower than that of the other two countries. So Fig. 8.11 would imply a declining market share for the Russians in the diminishing import mix. The output of radiata pine is clearly not going to be a limiting factor in this area and imports from Russia will slowly be supplanted as economic growth diverts their supply to the domestic economy. Thus, rather neatly perhaps, the forecasts here could well fit into the analysis of the Russian market in Chapter six.

An Overall Picture of the Coniferous Timber Market

Whilst the most interesting forecasts generated by the gentan probability model are those which are to be found for the individual sub-sectors of the coniferous timber market it is useful to amalgamate the forecasts. The following figures are formed by summing the forecasts from the various models. Thus, no interaction is assumed. The results are shown in Table 8.6.

Table 8.6. The market for coniferous timber

	Domestic output		Imports		Total demand	
	2005	2015	2005	2015	2005	2015
Base case	18,996	18,888	23,005	22,029	42,001	40,916
High GDP growth	18,146	18,112	25,493	26,628	43,640	44,739
Low GDP growth	19,624	19,552	19,805	17,025	39,492	36,577
Low GDP and log function	22,681	24,104	16,748	12,473	39,492	36,577
Low GDP and extreme case	25,460	31,394	13,929	5,051	39,389	36,445

All figures in thousands of cubic metres RWE.

The base case is that which contains the exogenous variables in the tables above. In the 'high GDP growth' case the higher levels of economic growth and the lower growth rates of North American timber prices were assumed, making it a pessimistic forecast from the point of view of Japanese output. In the 'low GDP growth' case the higher growth rates of American timber prices in Table 8.4 were used. In the case called 'low GDP log function' the form of the function describing the change in the parameters of the gentan probability distribution for sugi used the logarithm of time as a variable. Also the low growth rates of GDP and high log price growths were assumed. In the 'low GDP extreme function' case the assumption was made that the felling probabilities for sugi, hinoki and the other conifers remain the same into the future, the GDP growth rate is low but *beizai* prices are high. This is the scenario which most favours Japanese output.

The overall effect on the self-sufficiency ratio is shown in Fig. 8.12 for the 'high GDP growth' case and the 'low GDP extreme function' case. These represent the upper and lower limits on the trend in the ratio for the shown cases in Table 8.6. The graph makes it clear just how sensitive the forecasts for Japanese output, and therefore imports, are to what are only relatively slight changes in the parameters of two equations in the forecasting system.

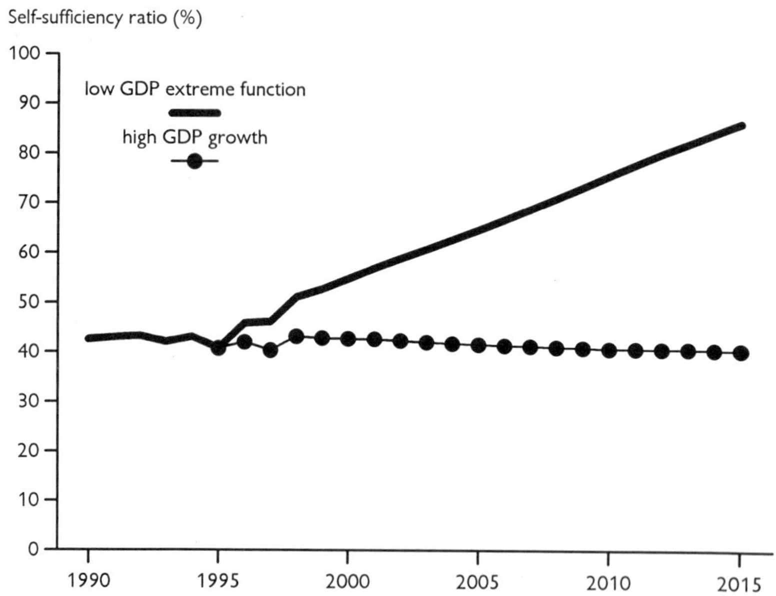

Fig. 8.12. The self-sufficiency ratio for coniferous timber

Plywood Demand

The models just presented represent the main forecasting effort here. However, a plywood model will be offered as it illustrates the depth of the market for panel products in general. The market for other panels, as was described in Chapter four, is one which is less developed in Japan than in other countries and it seems clear that the demand for such products will increase in the future and the increasing demand for *bunjou* housing shown in Fig. 8.5 gives some indication of this potential. However, given the relative shortness of the time series in this area, and the sensitivity of the supply side of the Japanese forestry system as just illustrated, only a simple model of plywood demand will be offered here.

The demand for plywood was found to be explicable by the area of *bunjou* housing starts, the proportion of wooden housing and the prices of imported plywood and substitute materials, as shown in Table 8.15 in the appendix. The forecasts of plywood demand are shown in Fig. 8.13 for the base scenario, as well as for the high GDP/low *beizai* price and low GDP/high *beizai* price cases.

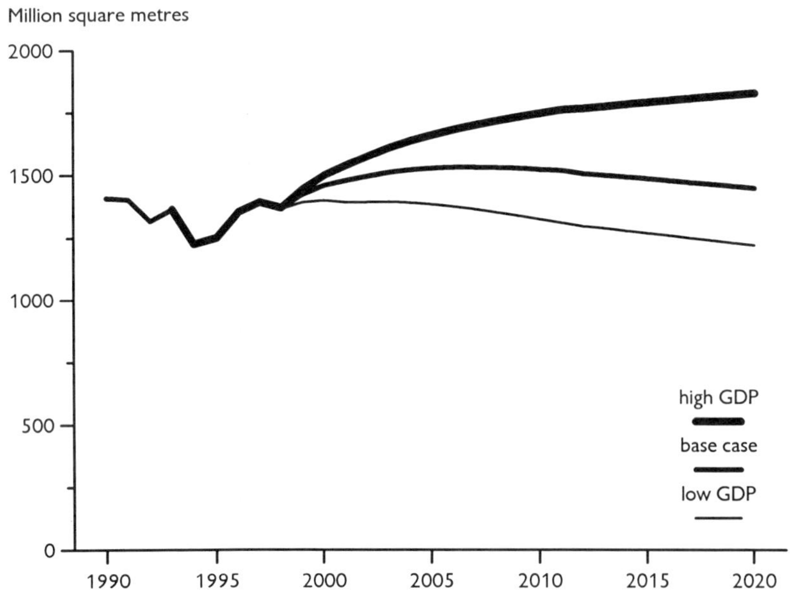

Fig. 8.13. The demand for plywood

An increase in the level of economic activity measured by GDP allows an increase in housing starts and an increase in the proportion of wooden housing which together increase the demand for plywood. Similarly, the demand for *bunjou* housing grows and this causes the demand for plywood to rise. A low price for North American timber imports stimulates the amount of wooden housing and also increases the proportion of *bunjou* housing in the overall mix. These factors account for the differences in the forecasts presented.

The proportion of this total demand that is imported is likely to depend very much on the policies of the various countries which, in the past, have supplied logs to Japan. The figures here might best be thought of as being indicative of the panel products market as a whole. The overall level of demand is likely to be relatively stable as the housing market and general economic activity are not showing any major upward or downward trends. However, the composition of the total demand figures shown in Fig. 8.13 is going to depend very much on relative rather than absolute prices and the marketing efforts of those involved in the production of the various panels which, technically at least, can substitute for plywood. This has been discussed in previous chapters.

Conclusions

The overall conclusion here is quite straightforward but, sadly, not very satisfactory. The predicted output of Japanese forests is very sensitive to the assumptions and, therefore, this sensitivity is reflected in the demand for imports. The linear model of the evolution of the parameters of the gentan probability distribution implies that the reduction in management intensity will continue. It produces forecasts of gentle increases in output. The logarithmic or extreme cases, which imply a stabilization of the situation in the rural economy, give rise to rapidly increasing outputs.

If anything, the linear model seems to be the more unreasonable of the two sets of equations describing the evolution of the gentan probability distribution. For this equation to be true the mean of the felling age must continue to increase into the future at the same rate as it has since 1965. But, most of the trends in the rural economy seem to have played themselves out. The increase in alternative employment has already made large inroads into rural employment, the younger population has already migrated to the towns. A factor which might be pressed into service to support the idea that the intensity of management will continue to fall is an increased demand for the environmental and recreational services of forests. Alternatively, if it is argued that no one will enter the forestry profession, the ageing of the work force shown in Fig. 1.5 in Chapter one would signal the imminent demise of Japanese forestry as the present generation of workers retire.

Statistically, there is little to choose between the different formulations and so standard statistical analysis is of little use here. This, it has to be said, is partly the result of the poor nature of the data. The problems in deriving the parameters of the distribution which gives rise to the felling probabilities mean that the available data are very limited and of low quality. Sadly, a lot of the data which the Japanese collected, ostensibly for the purpose of estimating the parameters of the gentan probability distribution, is useless and the data which are available give only very inaccurate estimates (Blandon, 1991b). However, it is precisely here where some precision is needed.

It would appear that, if those making the investment decisions on behalf of the Japanese sawmilling industry are being rational, they expect the right-hand side of Fig. 8.8, or something like it, to emerge as there seems to be no rush to invest in anticipation of a burgeoning domestic output. Similarly, most Japanese experts seem to be of the opinion that no output will be forthcoming from Japan's forests in the near future.

However, the weaknesses of the model aside, the strength is that is exposes the implicit assumptions in the forecasts of output from Japan's forests. Although most Japanese seem to be pessimistic about the ability of Japan's forests to supply the market, the implication is that trends in the rural economy must accelerate rather than remain the same. Assuming that nothing will change into the future implies that output will increase.

Appendix

The following are the equations referred to in the text. Figures in italics are significant at the 5% level, those in bold at the 1% level. Dummy variables were often used to account for changes in the data around the oil shocks. These are not reported in the tables below. The equations below apply to the forecasting period. Some equations were fitted using ρ-differences and a Hildreth–Lu grid search procedure. Durbin–Watson (D.W.) statistics are also given.

Table 8.7. The parameters of the gentan probability distribution

		Independent variables				
Species	Dependent variable	Coefficient	Time	Stumpage price	R^2	D.W.
Sugi	Mean	*−1445.4*	**0.742**	*−0.00028*	0.91	1.40
	Standard deviation	*2090.2*	**1.054**	**0.0069**	0.92	1.13
Hinoki	Mean	*−1382.7*	**0.711**	*−0.00015*	0.90	2.03
	Standard deviation	*−1081.6*	*0.547*	0.00018	0.82	2.19
Sugi (log time)	Mean	*−43.0*	19.750	−0.00047	0.81	1.33
	Standard deviation	*−68.4*	21.520	0.000104	0.91	1.57

Stumpage prices in yen/cubic metre, time as calendar year or log of year minus 1945.

Table 8.8. The area of housing starts

	Coefficient	Real GDP	Real land price	Capital employed	R^2	D.W.
Mochi-ie	*74,339*	**0.0024**	**−199.0**	−1.55	0.86	1.91
Bunjou	*3,877*	**0.0110**	*−86.4*	*−1.75*	0.55	1.24
Other	*4,739*	−748.40		3.35	0.18	1.77

Real land price index equals 100 for 1960, capital employed was found by multiplying the land price index by the interest rate. Real GDP entered the equation for 'other' housing as the natural logarithm. All the dependent variables were in thousands of square metres.

Table 8.9. The proportion of wooden housing

	Coefficient	Log price	Wages	Substitute price	GDP	R^2	D.W.
Mochi-ie	*0.903*	**0.00000017**	**−0.00382**	**0.00075**		0.97	1.55
Bunjou	0.173	*−0.0000082*				0.34	1.64
Other	*0.08*	**−0.00000068**			**−2.4E−010**	0.77	1.70

Log prices were the average of sugi and *beizai* prices in yen. Wages and the price of substitutes were in index form. The dependent variables were all in the form of proportions. Only 'other' housing used GDP as an independent variable.

Table 8.10. The relationship between sugi and *beizai* log prices

	Coefficient	*Bunjou* ratio	R^2	D.W.
Sugi/*beizai* log price ratio	***1.20***	***–1.50***	0.69	1.20

Bunjou ratio defined in terms of floor area.

Table 8.11. The demand for sugi and *beizai*

	Coefficient	Average log price	Housing starts	Substitute price	R^2	D.W.
Demand for sugi and *beziai*	6630	***–0.613***	***0.130***	*75.3*	0.56	2.06

Beizai figures in RWE. Substitute prices in index form.

Table 8.12. The ratio of logs to lumber in *beizai* imports

	Coefficient	2 by 4 ratio	Exchange rate	Lagged ratio	R^2	Durbin's *h*
Ratio of logs to lumber	*0.28*	*0.99*	***–0.00032***	***0.443***	0.98	0.064

Exchange rate in terms of yen per US dollar.

Table 8.13. The demand for hinoki

	Coefficient	Hinoki/sugi log price	*Mochi-ie* housing starts	2 by 4 ratio	R^2	D.W.
Hinoki demand	*5,949*	***–117.3***	***0.01***	*–67.43*	0.94	1.85

Demand in thousands of cubic metres. *Mochi-ie* housing starts in area terms.

Table 8.14. Demand for other conifers

	Coefficient	Housing starts	*Matsu* log price	GDP	R^2	D.W.
Demand for other conifers	*9,774*	***0.179***	***–0.395***	***–0.0028***	0.77	1.73

Housing starts in area terms, 'other' housing only.

Table 8.15. The demand for plywood

	Coefficient	*Bunjou* area	Price index	Substitute price	R^2	D.W.
Plywood demand	569,076	27.34	–29.08	4,647	0.58	2.12

Demand in thousands of square metres.

Chapter nine:

Concluding Remarks and Afterword

In most forecasting exercises it is a relatively simple matter to produce a prediction that is an extrapolation of the existing time series under examination. Indeed, it is often the case that a great deal of statistical effort is expended to produce a forecast that could have been produced merely by extending the time series into the future by eye. However, in the case of Japanese forestry the opposite is the case. Producing a forecast that follows the trends which have been seen up to now is actually very difficult, and the easiest forecasts are those which create an upward trend in domestic output and falls in imports. This leaves the forecaster with rather an exposed feeling of predicting something which many observers believe will not materialize.

Similarly, the actions of the participants in the market are such that it appears that there is no great belief that an era of self-sufficiency is on the horizon. So the forecasts of increasing supply seem difficult to justify, even if the models predict them. What is clear, however, is that the key to the various forecasts lies in the behaviour of Japanese forest owners and that this will have a ripple effect onto the forest economies of most of the supplying countries in the world, as well as those countries which compete with Japan for timber products.

It is rather surprising that so little work has been done in this area in Japan. With one or two notable exceptions, modelling of output is not very well developed in Japanese forestry circles. Indeed, the overwhelming impression that the writer has gained is that the Japanese are very strong at examining the minutiae of various markets but relatively poor at putting this together to produce wide-ranging models and generalizations.

The result is that a wide range of possible figures can be produced and a summary of some of the main ones is shown in Table 9.1. Here the forecasts from the previous chapter, taken from various tables there, are compared with the forecasts of MAFF and those produced for the Asia-Pacific Forestry Sector Outlook Study (Zhang *et al.*, 1997). The figures for North American imports

Table 9.1. A comparison of the forecasts

	FAO		MAFF		Chapter eight	
	Sawnwood	Roundwood	2005	2015	2005	2015
North America	6.0	16.2	21.3	20.0	21.3	24.1
					11.8	9.3
Russia, New Zealand & Chile	0.9	29.4	5.7	4.5	5.0	3.3
Tropical timber	0.6	14.2	1.5	1.0		
Japan		31.3	24.2	32.0	25.5	31.4
			14.4	14.0	19.0	18.8

Figures in millions of cubic metres, MAFF and Chapter eight figures in RWE. FAO forecasts refer to 2010. MAFF figures taken from the optimistic forecast except for Japanese production where both sets of figures are presented.
sources: Zhang *et al.* (1997), *Shinrin kihon keikakau kenkyuukai* (1998)

range from around the 22 million cubic metre mark for the main forecasts to as low as just over nine in the most extreme case of the forecasts in Chapter eight. Thus, it is tempting to suggest that imports are unlikely to exceed the upper figures but could, if the Japanese rural sector develops in the correct way, be lower. The figures for imports from other countries are around the 4-5 million cubic metre level, except for the FAO forecasts which seems to favour log imports. However, the model in Chapter eight predicts a lower overall demand level and so declining imports.

The figures for Japanese output from the FAO study are for total industrial roundwood. In the MAFF case the figures show output from man-made forests and in the forecasts taken from the previous chapter they refer to all conifers. The range of outputs in the MAFF cases is clear and, for comparison, their total output forecasts for the optimistic case are slightly above those from FAO. The range of figures from Chapter eight shows the effects of the different assumptions on output. The maximum figures generated by the forecasts in Chapter eight are similar to those from MAFF but the lower ones are in excess of their forecasts showing the difficulty in simulating a continuing downward trend in the face of the maturing timber stock.

Apart from these observations on the forecasts examined here, many of the countries referred to in the text have experienced large changes during the preparation of this book. For example, the economies of Southeast Asia have 'collapsed' and the devaluation of the rouble has turned attention to the situation in Russia. Financial institutions in Japan are in a parlous state and this is affecting domestic confidence. All of these factors will have effects on the nature of the various countries' forest economies and, therefore, the forecasts contained here.

Of course, there are two possible interpretations that could be placed on such events. One is that they are temporary, 'one-off' events which will be seen in hindsight as being mere 'blips' in an otherwise continuing long-term trend of production and consumption. In this case, as the MAFF forecasts and the forecasts of the author are intended to be long-term ones, there is not much need to comment further except to make the point that the prognoses for the last years of the decade are likely to be inaccurate. On the other hand, these events could represent breaks in the trends observed in the data used to estimate the models. Thus, the data to hand would not allow the explicit incorporation of such factors and so the events in the recent history of the countries involved will cause the forecasts to be badly inaccurate.

Looking at the events in turn, the slowing of the Japanese economy presents a number of problems in interpretation. The collapse of the financial system has caused many commentators to argue that the economy in Japan should be opened up and made more like western ones, usually the American one. This is somewhat paradoxical as it was the movement of the Japanese away from what they did well, the production of products, into financial 'engineering' which could be argued lies at the root of the present problems. Already some Japanese commentators are worrying about the large downturn in the demand for housing and, therefore, timber products which the recession there is causing. In Chapter three is was suggested that large temporary changes in the demand for housing were possible as a result of short-term deferral of investment. However, some now seem to feel that the recent decline is not a temporary 'blip' but represents a long-term change in the trends in housing demand such as that which occurred after the first oil shock.

However, one factor, which seems likely to occur, is the movement of Japanese companies to restructure, indeed this was a trend that was becoming apparent before recent events. Whilst this might seem irrelevant for the forestry sector it is, in the writer's opinion, likely to be highly significant for a number of reasons. The bulk of the demand for housing, and therefore timber, has come from the savings of the salaried staff who have been more-or-less guaranteed a tenured job. These jobs have included large bonuses which allow the savings needed to invest in housing. But the famous Japanese paternalistic form of management has generated a high fixed-cost base for large areas of Japanese industry. If the cost cutting, which the recession in Japan seems to be creating, causes a move to a 'free' labour market, so that the security of the salaried staff is reduced, the effect will be felt in the housing market. Thus, the trend away from the more expensive *mochi-ie*, traditional types of housing towards *bunjou*, two-by-four types or the non-wooden mansions will be accelerated. As such the recent drop in Japanese growth rates would reinforce the trends already apparent with the younger Japanese preferring western-style housing and, therefore, creating more of a 'normal' market for softwood timber.

If this is the case, the forecasts of the housing starts in Chapter two, which it was suggested might be pessimistic, could well materialize and the forecasts of MAFF and those in the author's model in Chapter eight will become over-estimates. Not only that, but the type of timber used in the housing will probably tend towards the cheaper and the use of panels will increase. A greater proportion of the housing is likely to be provided by the large building firms rather than the smaller carpenters who specialize in the more traditional forms of building. Usually, however, it is the larger companies in this field that import a lot of the timber for their housing, as they are mainly building either large mansions of operating in the *bunjou* area. Thus, cost considerations are more important than they are for the smaller scale providers of *mochi-ie* types of housing. The clear implication is that the outputs of domestic timber will be lower than those suggested earlier, although in the author's model the drop in price would probably not be enough totally to choke off the impending increase in supply.

On the other hand, the Japanese government has traditionally used the construction and housing market as a vehicle to expand the economy. Already, calls for increased investment in these areas have been made and this could go some way to counter the lack of consumer confidence. So the recession could, paradoxically, cause an increase in the demand for housing. An indicator which might be worth observing here is expenditure in 'do-it-yourself' stores. As was noted in Chapter two, the growth in DIY which has been seen in Britain has not really been reproduced in Japan, with only a small number of such outlets in existence. An increase in expenditure on house maintenance could presage a change in the attitude of the Japanese to their housing from being a short term item to a more permanent consumer durable.

Another effect of the Japanese recession is that many Japanese firms are thinking of returning home. During the years of the so-called 'bubble economy' a number of firms moved their production operations to Southeast Asia for cost reasons, causing worries of a 'hollowing out' of the Japanese economy. The return to a more inward-looking investment policy by the Japanese will mean that co-operative ventures with other countries will be much reduced, or if not reduced then changed in form. The effect of the recession is likely to be a preference for foreign investment to return to its previous pattern of securing supplies of raw material rather than investment in manufacturing capacity to produced more processed products. Thus, it might be expected that the recession will create greater log imports so breaking the trends towards processed product imports seen until now, especially from Southeast Asia and Russia.

One factor which might accentuate this is a reduction in environmental considerations in Japan. It is usually the case that, in the main, the sacrificing of consumption to pursue environmental concerns is stronger when economies are growing. The environmental movement in Japan has always been weaker than in other countries, with issues in this area not being so fully aired or considered by the general population. Thus, the moves to coniferous plywood, in as far as they are not driven by purely economic motives, will be reduced.

Of course, the countries supplying tropical timber to Japan have also seen major economic upheavals during the preparation of this book. The comments about the environmental movement apply equally to them. The drop in the value of their currencies and the collapse of markets are likely to cause an attempt to maximize foreign exchange earnings. The easiest way to do this is by the export of raw materials such as logs or tropical hardwood based plywood. Thus, it is quite possible that the forecasts of MAFF for the imports of tropical hardwood logs are underestimates. Also, countries such as Indonesia will view their agricultural sectors as being somewhat insulated from international markets and so encourage their further development. This will also help to diminish some of the effects of unemployment in the outward-oriented export sector. The result will be to put further pressure on the forestry sector and could lead to a reduction in attempts to manage forests in a sustainable way. On the other hand, large investment projects, such as roading in Kalimantan, which might have been expected to increase log outputs could well be cut back in the present circumstances.

If this is the case it will have an effect on other panel products, for the dominance of tropical plywood in the Japanese panels market is largely based on its relative price. A drive to maximize foreign exchange earnings by exports, especially as the devaluation of their currency will reduce the price of such plywood, could well reduce the need of the Japanese to move to other panel products and this will be felt in the American markets. Indeed, the price of tropical plywood fell dramatically during the preparation of this book and has caused many of the producers to come close to bankruptcy.

The recent collapse in the value of the rouble is, in one sense, a catching up of the financial sector with what had already happened in the production sector. However, the effect is going to be a reduction in foreign investment in the Russian economy and, as Chapter six showed, this is likely to reduce the available growing stock in the future and the long-term ability of the Russians to maintain a timber export presence. However, the reduction in domestic economic activity reduces internal demand and, like the countries in Southeast Asia, the domestic difficulties will generate a demand for foreign exchange to be obtained in the traditional way by exporting raw materials. So the exports of logs from Russia are likely to be less effected than any potential development into processed products.

Against this background is the recent drop in the value of the yen with respect to the US dollar and the problems of the Korean economy. The first will mean an increase in the price of North American forest products within Japan, implying that the high *beizai* log price growth scenarios in the previous chapters would more closely simulate events. The second would suggest that, despite its problems, Japan will strengthen its relative competitive position in the Pacific Rim timber trade. This will increase domestic production and reduce imports. So one is tempted to suggest somewhat lower imports from North America, greater ones from Southeast Asia and a slowing of trends towards panels, especially from Southeast Asia.

What is clear is that, from the point of view of Japan's role in the world timber markets, the future is unlikely to be an extrapolation of the recent past.

Glossary of Japanese Terms

This glossary contains brief definitions or descriptions of the Japanese words that are used in the main text. Although the Japanese and English names for the main tree species were presented in Table 1.3 of Chapter one, for the sake of completeness they are also listed here. In the main text Japanese words, with the exception of the names of the tree species and those words in normal English usage, were italicized. This convention is maintained here. The words in brackets at the end of some of the entries are the definitions given in *Vocabulary of forest science*, which was published in 1990 by the Japan Forestry Foundation.

Akamatsu. Red pine, *Pinus densiflora*.

Beizai. Timber deriving from North America. (US timber, North American wood.)

Buna. Beech, *Fagus crenata*.

Bunjou. A term applied to housing construction that refers to houses built in lots and then sold separately after completion. Usually, this is the preserve of larger companies who purchase land and then build houses on it prior to finding buyers.

Chuu maruta. Literally this means middle-sized logs. The actual dimensions referred to differ from species to species. For example, in the case of sugi and hinoki the diameter of the logs would be 14 to 22 cm and the length 3.65 to 4 metres. In the case of pine, however, the diameter of *chuu maruta* would be 24 to 28 cm.

Doboku kenchiku. This term means civil engineering works. In this context it refers to timber used for any building end-use except the construction of accommodation.

Ezomatsu. *Picea jezoensis*.

Gentan. Literally this means reduction in area but it is the name given to the probability that a forest, planted now, will be felled at some time in the future. In normal usage, this word is used in agriculture to refer to the policy of reducing the area under rice cultivation in Japan.

Hikidoritsu. An uncommon word but used in building data to mean the proportion of houses with sliding doors for the front entrance. This is taken as a characteristic of Japanese-style houses.

Himokuzou juutaku. Literally this means housing not made of wood. In the context here it refers to housing that is classified as 'non-wooden' implying that the pillars, sills, beams or staircases are not made of wood.

Hinoki. Japanese cypress. *Chamaecyparis obtusa*.

Hi-noukarinka. This is a forest owning household that is defined in the Japanese census as not being an agricultural family. That is to say, a forest owning household whose main source of income is not from agriculture. (Non-farm household forest owner, non-farmer forest owner.)

Hokkaido. The main northern island in the Japanese archipelago.

Hokuyouzai. In the context here this term means timber or wood from Russia or, formerly, the Soviet Union. (Boreal timber.)

Honshu. The central island in the Japanese archipelago.

Houryuu-ji. A famous temple in central Japan. Reputed to be the oldest wooden building in the world.

Iyokuteki. Literally this means 'highly motivated' or 'enthusiastic'. Here it is applied to one of the latest forecasts produced by the Japanese and implies that the forecasts will not materialize unless the accompanying policies are rigorously applied.

Jikyuuritsu. This means self-sufficiency ratio or the proportion of a certain category of goods that is supplied domestically.

Jinkourin. Man-made forest .

Jyuutaku kenchiku. Housing construction.

Kakudai zourin. Literally this means afforestation to expand the area of forest, but in the Japanese context the meaning is usually the replacement of a natural forest with a man-made one. (Expansive afforestation.)

Karamatsu. Japanese larch. *Larix leptolepis*.

Kashi-ie. A rented house or housing produced specifically for the rental market.

Kokusanzaijidai. This term means an era of self-sufficiency or, more literally, an era of domestic production. This term is used to refer to the idea that the maturing Japanese timber stock will be felled and so create a period when timber demand can be satisfied from domestic sources.

Kokuyuurin. This is national forest, that is forest managed by central government. (National forest.)

Konbouyou. Timber used for packaging, etc.

Kon-pane. Plywood, often of tropical origin, used as concrete forming. It is derived from the words 'concrete' and 'panel'.

Kuromatsu. Black pine. *Pinus thunbergii*.

Kyushu. The main southern island of the Japanese archipelago.

Kyuu-you. Housing supplied by a firm or organization for its employees.

Migaki-maruta. These are specially produced logs intended for use as decorative pillars to be used as a feature in traditional Japanese rooms. (Polished log.)

Minyuurin. This term refers to forest that is privately or publicly owned. It excludes national forest but includes forests owned by towns, prefectures, individuals and companies. (Private forest, non-national forest.)

Mitooshi. The dictionary definition of this word is usually something like 'perspective' or 'unobstructed view'. In the context here it is the word used to describe the forecasts or projections of timber supply and demand produced by the Japanese government.

Mochi-ie. Houses or housing produced for owner-occupiers. These houses are usually produced for specific owners, often on an individual basis.

Mokuzou. Made of wood.

Mokuzouritsu. In the context of this book the word refers to the proportion of housing that is defined as 'wooden', i.e. the pillars, beams, sills and staircases are wooden. The proportion of wooden housing is defined in two ways: either it is found by dividing the number of wooden housing starts by the total number of starts in a given period; or else figures for the floor area are used.

Momi. *Abies firma.*

Nanyouzai. Timber from tropical countries, usually Southeast Asia. (Tropical timber.)

Nara. Oak. *Quercus* spp.

Noukarinka. This is a term used to define forest owners who are also categorized as farmers. (Farm household forest owner.)

Purekatto. Derived from the Japanese pronunciation of 'precut' this term is applied to housing produced by cutting the timber to size in a factory and then shipped to the site to be assembled. It differs from prefabricated housing in that CAD/CAM systems are used to produce individually designed houses.

Sakoku. The name of the policy enforced before the Meiji Restoration that shut the country off from the rest of the world.

Seizaiyouzai. Saw timber, logs used for sawing.

Shichousonzaisanku. This term applies to forests owned by any non-private group that is not national or prefectural, for example cities, towns and villages. (Municipal forests.)

Shiitake. A type of mushroom. (*Lentinus edodes,* shii-take mushroom.)

Shikoku. The fourth largest island in the Japanese archipelago.

Shinsetsu jyuutaku chakkou kosuu. Housing starts. This term refers to measures of housing starts in terms of number of starts rather than floor area.

Shiyuurin. Private forest. This category of forest includes forest owned by individuals, companies, temples, etc. (Private forest.)

Sugi. Japanese cedar. *Cryptomeria japonica.*

Suusei. The dictionary definition of this word is 'tendency' or 'drift'. Here it is applied to one of the latest forecasts that, it is argued, will materialize if the present trends continue. It has been translated in this book as 'extrapolative'.

Tennenrin. Literally this means natural forest but in most contexts would not imply untouched or virgin stands. In the Japanese context the term usually implies any forest that is regenerated by means other than planting of seedlings or cuttings. (Natural forest, primeval forest.)

Todomatsu. *Abies sachalinensis.*

Todoufuken. Forests owned by prefectures.

Tsuga. *Tsuga sieboldii.*

Washitsu. A Japanese-style room, usually one with exposed pillars.

Youzai. Industrial wood, wood used for anything except firewood or *shiitake* bed logs. (Saw timber, log, industrial round wood.)

References

In the main text it has been the convention to italicize transliterated Japanese words. In the reference list this usage will not be pursued. Rather the usual convention of italicizing book and journal titles will be followed. In the text references to unattributed articles in journals such as *Mokuzai Jouhou* are referenced simply by month and year. They do not appear in this list.

Adams, D. M. and Haynes, R. W. (1989) Changing timber policies may bring other changes. *Journal of Forestry* 87, no. 4, 24-31.

Akaha, T. (1992) *Nihon ringyou no seisan kouzou*. Nourin toukei kyoukai, Tokyo.

Alig, R. J. and Wear, D. N. (1992) Changes in private timberland. *Journal of Forestry* 90, no. 5, 31-36.

Amano, M. and Noda, I. (undated) *Basic plan regarding forest resources in Japan*, mimeo.

Amano, M, Noda, I. and Kumazaki, M. (1984) Timber supply projection model for private forests in Japan. In: *Proceedings of IUFRO Symposium on Forest Management and Managerial Economics*. Tokyo University, pp. 285-292.

APRENet (1994) http://www.nautilus.org/aprenet/regional/susdevRus.html

APRENet (1997) http://www.nautilus.org/aprenet/library/tumen/forest.html

Araya, A. (1998) Heisei 9 nen no mokuzai yunyuu jisseki. *Mokuzai Jouhou*, no. 6, 5-12.

Azuma, T. (1995) Gouban seisan no doukou. *Journal of Forest Economics*, no. 127, 47-52.

Backman, C. A. (1996) *The Russian forest sector: opportunities and challenges.* Working Paper WP-96-53, International Institute for Applied Systems Analysis, Laxenburg, Austria.

Backman, C. A. and Waggener, T. R. (1994) *The Russian forestry sector outlook and export potential for unprocessed logs and primary forest products through 2000*. Working Paper 46, Center for International Trade in Forest Products, University of Washington, Seattle, Washington.

Ball, K. W. (1997) *Japanese trade and consumption of tropical timber resources*, unpublished MA thesis, University of Limerick, Ireland (http://forests.org/gopher/japan/japadiss.txt).

Barbier, E. B., Burgess, J. C., Bishop, J. and Aylward, B. (1994) *The economics of the tropical timber trade*. Earthscan, London.

Barnes, C. (1998) Hokubei ni okeru OSB jyukyuu no genjou. *Mokuzai Jouhou*, no. 2, 1-9.

Barr, B. M. and Braden, K. E. (1988) *The disappearing Russian forest*. Rowman and Littlefield, New Jersey.

Beasley, W. G. (1982) *The modern history of Japan*. Charles E. Tuttle, Tokyo.

Blandon, P. R. (1983) *Soviet forest industries*, Westview Press, Boulder, Colorado, USA.

Blandon, P. R. (1991a) *Studies on gentan probability theory and the felling behaviour of Japanese private forest owners*. The University of Tskububa, Tsukuba, Japan.

Blandon, P. R. (1991b) Censored sample theory and the forecasting of output from Japanese private forests (I). *Journal of the Japanese Forestry Society* 73, no. 5, 187-196.

Blandon, P. R. (1993) Unmanaged forests and forecasting using the gentan probability distribution. *Journal of the Japanese Forestry Society* 75, no. 6, 484-492.

Blandon, P. R. (1994) Movements in the parameters of the gentan probability distribution. *Journal of the Japanese Forestry Society* 76, no. 1, 1-9.

Blandon, P. R. (1996) *Igirisujin ga mita nihon ringyou no shourai*. Tsukijishoukan, Tokyo.

Briggs, D. G. and Buhler, C. (1989) Pacific Northwest hardwoods in international trade. *Forest Products Journal* 39, no. 7/8, 28-33.

Brooks, D. J. (1995) Federal timber supply reductions in the Pacific Northwest. *Journal of Forestry* 93, no. 7, 29-33.

Brown, C. (1997) *In depth country study - New Zealand*. Asia-Pacific Forestry Sector Outlook Study Working Paper APFSOS/WP/05, FAO, Rome.

Cardellichio, P., Youn, Y. and Adams, D. (1989) *A preliminary analysis of timber and timber products production, consumption, trade and prices in the Pacific rim until 2000*. Working Paper 22, Center for International Trade in Forest Products, University of Washington, Seattle, Washington.

Cardellichio, P. A., Binkley, C. S. and Zausaev, V. K. (1990) Sawlog exports from the Soviet Far East. *Journal of Forestry* 88, no. 6, 12-17 & 36.

Cengel, D. and McKillop, W. (1990) U.S. trade in transition - an econometric view of south sea wood flows. *Forest Science* 36, no. 2, 425-437.

Chelyshev, V. and Abe, N. (1997) Condition and dynamics of the Russian Far East forests in 1966-1993. *Journal of Forest Planning* 3, no. 1, 31-40.

Dauvergne, P. (1997) *Shadows in the forest, Japan and the politics of timber in*

Southeast Asia. MIT Press, Cambridge, Massachusetts.

Dennis, D. F. (1989) An economic analysis of harvest behavior: integrating forest and ownership characteristics. *Forest Science* 34, no. 4, 1088-1104.

Doki, Y. (1994) Hitsuyou na youto ni oujita jushu no sentei. *Mokuzai Jouhou*, no. 12, 14-16.

Elwood, N. E. (1992) Forestry cooperatives in Japan. *Journal of Forestry* 90, no. 6, 25-28.

FAO (1997) *The state of the world's forests*. Food and Agricultural Organization, Rome.

FAO (1998) *Japan - in depth country study*. Asia Pacific Forestry Outlook Study Working Paper APFSOS/WP/15, Food and Agricultural Organization, Rome.

FICGB (1996) *The forestry industry yearbook*. Forest Industry Council of Great Britain, London.

Fujimori, T. (1985) *Edauchi to ikurin gijutsu*. Nihon Ringyou Chousakai, Tokyo.

Funakoshi, S. (1981) Historial development of forest policy in Japan. In: Handa, R. *et al.* (eds) *The Current State of Japanese Forestry I*. The Japanese Forest Economics Society, Tokyo, pp. 112-117.

Funakoshi, S. (1988) The process and keynote of forest policy in the pre-world II era. In: Handa, R. (ed.), *Forest Policy in Japan*. Nippon Ringyou Chousakai, Tokyo, pp. 16-21.

Gottko, J. and McMahon, R. (1989) Oregon lumber producers: evaluating exporting in the 1980s. *Forest Products Journal* 39, no. 11/12, 29-32.

Grainger, A. (1990) Modelling deforestation in the humid tropics. In: Palo, M. and Mery, G. (eds.), *Deforestation or development in the third world*. Bulletin 349 of the Finnish Forest Research Institute, Helsinki, pp. 51-67.

Granåsen, J, Nilsson, S. and Zackrisson, U. (1997) *Russian forest sector - human resources*. Interim report IR-97-008/March, International Institute for Applied Systems Analysis, Laxenburg, Austria.

Gwynne, R. N. (1993) Non-traditional export growth and economic development: the Chilean forestry sector since 1974. *Bulletin of Latin American Research* 12, no. 2, 147-169.

Handa, R. (1988) Timber economy and forest policy after world war II. In: Handa, R. (ed.), *Forest Policy in Japan*. Nippon Ringyou Chousakai, Tokyo, pp. 22-36.

Handa, R. (1990) *Rinseigaku*. Buneidou, Tokyo, 1990.

Hansen, B. and Luppold, W. (1992) Factors affecting Japanese hardwood lumber imports, 1976-1990. *Forest Products Journal* 42, no. 9, 47-51.

Hashimoto, M. (1995) Yunyuu juutaku no genjou to tenbou. *Mokuzai Jouhou*, no. 8, 6-9.

Haynes, R. W. (1990) *An analysis of the timber situation in the United States*. USDA Forestry Service General Technical Report RM-199, US Department of Agriculture.

Haynes, R. W. and Adams, D. M. (1992) The timber situation in the United

States. *Journal of Forestry* 90, no. 5, 38-43.

Himeno, T. (1995) Mokushitsu bo-do no jyuyoukoudou. *Mokuzai Jouhou*, no. 11, 11-14.

Horgan, G. P. and Maplesden, F. M. (1995) The role of marketing in developing exports of plantation-grown New Zealand radiata pine. *Unasylva* 46, no. 183, (http://www.fao.org/WAICENT/FAOINFO/FORESTRY)

Iehara, T. (1993) The profitability and economic evaluation of the forest management of hinoki in a long-rotation operation. *Journal of the Japanese Forestry Society* 75, no. 1, 34-40.

Iida, S. (1992) *Kokuyuurin no kako genzai shourai*. Tsukuba Shobou, Tokyo.

Ishihara, T. (1980) *Shinrinkoujou e no michi*. Seibunsha, Tokyo.

Ishii, Y. and Arai, H. (1986) A study on the present state of plantation forestry in Hokkaido. In: Handa, R. *et al.* (eds.) *The Current State of Japanese Forestry V*. The Japanese Forest Economics Society, Tokyo, pp. 30-38.

ITTO (1997) *Annual review and assessment of the world tropical timber situation.* Document GI-7/97, International Tropical Timber Organization, Yokohama, Tokyo.

Izumi, E. (1988) A brief history of Japanese forests and forestry up to the 19th century. In: Handa, R. (ed.), *Forest Policy in Japan.* Nippon Ringyou Chousakai, Tokyo, pp. 151-160.

Jélvez, A., Blatner, K. A. and Govett, R. L. (1990) Forest management and production in Chile. *Journal of Forestry* 88, no. 3, 30-34.

JFTA (1982) *Wood supply and demand in Japan.* Japan Forest Technical Association, Tokyo.

Kakizawa, H. (1994) *Market potential for Russian Forest Products in Japan.* Working Paper WP-94-94, International Institute for Applied Systems Analysis, Laxenburg, Austria.

Kakizawa, H. (1998) Roshia kyokutou no shinrin kanri. *Mokuzai Jouhou*, no. 3, 7-16.

Kakizawa, H. and An, K. (1997) Forest policy reform and the restructuring of the forest industry in Russia. *Journal of Forest Research* 2, no. 1, 25-30.

Kallio, M., Dykstra, D. P. and Binkley, C. S. (eds) (1987) *The global forest sector: an analytical perspective*. John Wiley and Sons, New York.

Kata, Y. (1995) Hanshin awaji daishinsaigo no fukkoukeikaku to mokuzai ryuutsuu. *Mokuzai Jouhou*, no. 12, 13-15.

Kato, T. (1981) A regional comparison of forest productivity, stumpage price, logging and regeneration costs among Japan, Canada and the United States. In: Handa, R. *et al.* (eds) *The Current State of Japanese Forestry I.* The Japanese Forest Economics Society, Tokyo, pp. 29-40.

Kawata, I. (1992) The importation of New Zealand's wood and the packaging lumber industry. In: Handa, R. *et al.* (eds) *The Current State of Japanese Forestry VIII.* The Japanese Forest Economics Society, Tokyo, pp. 102-114.

Koide, Y. (1992) Industrial utilization of hardwood timber in Japan. In: Handa, R. *et al.* (eds) *The Current State of Japanese Forestry VIII.* The Japanese

Forest Economics Society, Tokyo, pp. 33-39.

Kondrashev, L. (1997a) Yuukou na shudan wa gaishi. *Mokuzai Jouhou*, no. 3, 23-25.

Kondrashev, L. (1997b) Roshiyakyokutou ringyou no jittai to tainichi yushutsu. *Mokuzai Jouhou*, no. 7, 1-7.

Kondrashev, L. (1998) Saikin Roshiya keizai to kyokutouringyou. *Mokuzai Jouhou*, no. 3, 17-21.

Konohira, Y. (1995) Nyuu Jirando ni okeru rajiataa pain no shijoudoukou to keiei. *Mokuzai Jouhou*, no. 6, 13-20.

Krankina, O. and Ethington, R. L. (1995) Forest resources and wood properties of commercial tree species in the Russian Far East. *Forest Products Journal* 45, no. 10, 44-50.

Kumazaki, M. (1997) Nihon kokuyuurin seisaku. *Sekai*, no. 2, 170-178.

Kumazaki, M. (1998) Kakakuhendou ni miru sugi hinoki ringyou no 40 nen. *Rinkeikyou geppou*, no. 442, 2-14.

Levina, L. I. (1997) Sakhalinskii les na Yaponskom rinkye. *Lesnaya Promishlennost'*, no. 2, 25-29.

Lipman, D. (1994) The Russian forest industry during the transitional period. *Unasylva* 45, no. 179, 19-24.

Lochkarev, B. A. (1995a) Lesnoi eksport: problemi effectivnosti. *Lesnaya Promisshlenost'*, no. 1, 5-6.

Lochkarev, B. A. (1995b) Lesniye tovari iz Rossii: nuzhni pravila torgovli. *Lesnaya Promisshlenost'*. no. 3, 5-6.

Loikkanen, H. A., Kuuluvainen, J., and Salo, J. (1986) *Timber supply of private nonindustrial forest owners: evidence from Finland.* Research Report no. 50, Institute of Economics, University of Helsinki, Helsinki, Finland.

MAFF (1980) *Basic plan for Japan's forest resources and long range demand and supply projection for important forest products.* Ministry of Agriculture, Forestry and Fisheries, Tokyo.

MAFF (1996) *Basic plan for Japan's forest resources and long range demand and supply projection for important forest products.* Ministry of Agriculture, Forestry and Fisheries, Tokyo.

Maplesden, F. (1993) Japanese sawmilling industry: current situation, historic trends and a comparison with the New Zealand industry. *New Zealand Journal of Forestry Science* 23, no. 2, 209-235.

Marcot, B. G. and Thomas, J. W. (1997) *Of spotted owls, old growth, and new policies: a history since the interagency scientific committee report.* Pacific Northwest Research Station General Technical Report, PNW-GTR-408, US Department of Agriculture.

Matsuo, M. (1995) Dariekkusupo-toresu ni yoru rainen no tainichi yushutsu. *Mokuzai Jouhou*, no. 12, 10-12.

MOFA (1997) http://www2.nttca.com:8010/infomofa/jr/russia_policy.html

Murashima, Y. (1996) Jishingo ichinen no mokuzai jyuyou no suii. *Mokuzai Jouhou*, no. 8, 6-9.

Nakamura, T. (1981) *The postwar Japanese economy.* University of Tokyo Press, Tokyo.

Narita, M. (1996) Forest management and forest labor. *Journal of Forest Planning* 2, no. 2, 97-105.

Narita, M. (1997) Forest management and forestry. *Journal of Forest Planning* 3, no. 1, 41-47.

Nectoux, F. (1989) *Timber from the south seas.* World Wide Fund for Nature International, Gland, Switzerland.

Nilsson, S. (ed.) (1997a) *Dialogue on sustainable development of the Russian forest sector - volume I.* Interim report IR-97-009/March, International Institute for Applied Systems Analysis, Laxenburg, Austria.

Nilsson, S. (ed.) (1997b) *Dialogue on sustainable development of the Russian forest sector - volume II.* Interim report IR-97-010/March, International Institute for Applied Systems Analysis, Laxenburg, Austria.

Nishikawa, K. (1986) Trends of silviculture and management of hardwood forests in Japan. In: Handa, R. *et al.* (eds) *The Current State of Japanese Forestry V.* The Japanese Forest Economics Society, Tokyo, pp. 19-29.

Noda, I. and Amano, M. (1983) Minyuurin ni okeru mokuzai kyoukyuu yosoku. In: *Proceedings of the 94th Conference of the Japanese Forestry Society*, Japanese Forestry Society, Tokyo, pp. 113-116.

Nomura, I. (1981) The current state of the plywood industry and the tropical hardwood import of Japan. In: Handa, R. *et al.* (eds) *The Current State of Japanese Forestry I.* The Japanese Forest Economics Society, Tokyo, pp. 21-28.

Oka, H. (1997) Classification and choices of management regimes for marginal timber plantations. *Journal of Forest Planning* 3, no. 2, 91-97.

O'Laughlin, J. (1992) What the law is and what it might become. *Journal of Forestry* 90, no. 8, 6-12.

Perez-Garcia, J. M. (1995) Global and landuse consequences of North American timberland withdrawals. *Journal of Forestry* 93, no. 7, 34-38.

Rinaco (1996) http://www.fe.msk.ru/infomarket/rinacoplus/newslett

Rinyachou (1984) *Kokuseizai kyoukyuu shisutemu keisan moderu kaihatsu chousa houkokusho.* Rinyachou Kikakuka, Tokyo.

Rinyachou (1997) Shinrin shigen ni kan suru kihonkeikaku narabi ni jyuuyou na rinsanbutsu no jyuyou oyobi kyoukyuu ni kan suru chouki no mitooshi. *Rinnou jihou*, no. 1, 5-19.

Sakaguchi, K. (1983) *Sugi no subete.* Zenkoku Ringyou Kairyou Fukyoukai, Tokyo.

Sakai, M. (1992) Changes in location of the sawmill industry in Japan in the 1980s. In: Handa, R. *et al.* (eds.) *The Current State of Japanese Forestry VIII.* The Japanese Forest Economics Society, Tokyo, pp. 40-61.

Sample, V. A. and Le Master, D. C. (1992) Economic effects of northern spotted owl protection. *Journal of Forestry*, 90, no. 8, 31-35.

Schallau, C. H. and Goetzl, A. (1992) Effects of constraining US timber supplies.

Journal of Forestry 90, no. 7, 23-27.

Schäffler, L (1998) Pruning - luxury or good business? *New Zealand Pine*, March-April (http://www.nzforestry.co.nz/html/nz_pine)

Sedjo, R. A. (1995) Local logging, global effects. *Journal of Forestry* 93, no. 7, 25-28.

Sedjo, R. A., Wiseman, A. C. and Lyon, K. S. (1992) Exploiting national returns to log exports. *Journal of Forestry* 90, no. 4, 35-39.

Sedjo, R. A., Wiseman, A. C., Brooks, D. J. and Lyon, K. S. (1994) *Changing timber supply and the Japanese market*. Discussion paper no. 94-25, Resources for the Future, Washington.

Shinrin Keikaku Kenkyuukai (1987) *Aratana shinrin ringyou no chouki bijyon*. Shinrin Keikaku Kenkyuukai, Tokyo.

Shinrin Keikaku Seidou Kenkyuukai (1992) *Shinrin keikaku no jitsumu*. Shinrin Keikaku Seidou Kenyuukai, Tokyo.

Shinrin Kihon Keikaku Kenkyuukai (1997) *21 seiki wo tenbou shita shinrin ringyou no chouki bijyon*. Shinrin Kihon Keikaku Kenkyuukai, Tokyo.

Shinrin Ringyou Gyousei Kenkyuukai (1988) *Shinrin ringyou mokuzaisangyou no kihon shihyou*. Shinrin Ringyou Gyousei Kenkyuukai, Tokyo.

Sugahara, S. and Ouchi, Y. (1988) Development of the forest plan system. In: Handa, R. (ed.), *Forest Policy in Japan*. Nippon Ringyou Chousakai, Tokyo, pp. 37-43.

Suzuki, T. (1972) Ringyou ni okeru kakuritsukateiron no ouyou I. *Journal of the Japanese Forestry Society* 54, no. 7, 234-237.

Suzuki, T. (1973) Ringyou ni okeru kakuritsukateiron no ouyou II. *Jounal of the Japanese Forestry Society* 55, no. 7, 234-237.

Suzuki, T. (1979) *Shinrin keirigaku*. Asakurashoten, Tokyo.

Suzuki, T. (1996) Shinyoujugouban no jyukyuudoukou to kadai. *Mokuzai Jouhou*, no. 6, 6-10.

Suzuki, T. (1997) Shutoken ni okeru sugi ippanzai seihin no jyuyou doukou. *Mokuzai Jouhou*, no. 6, 14-17.

Takaki, H. (1997) Shutokenkinkou purekatto kouhou no genba kara genjou to shouraitenbou. *Mokuzai Jouhou*, no. 9, 12-19.

Tanaka, S. (1997a) The view from Japan. *New Zealand Pine*, no. 15, March-April, (http://www.nzforestry.co.nz/html/nz_pine)

Tanaka, S. (1997b) Fuzaison shinrinshoyuusha to shinrinkumiai. *Shinrin kumiai*, no. 9, 14-18.

Totman, C. (1987) Lumber provisioning in early modern Japan. *Journal of Forest History* 31, no. 2, 56-70.

Tsukamoto, M., and Fujiwara, M. (1995) The structure of the hardwood log market - correlation analysis of quality and price on the basis of log buyers' bidding behaviors. *Journal of the Japanese Forestry Society* 77, no. 6, 507-515.

Ueki, T. (1992) The current situation and problem points of the change in policy to 'natural commercial forests' in national forests today. In: Handa, R. *et al.*

(eds) *The Current State of Japanese Forestry VIII*. The Japanese Forest Economics Society, Tokyo, pp.25-32.

UN-ECE/FAO (1996) *North American timber trends study*, ECE/TIM/SP/9, Geneva.

UN/ECE (1997) http://www.unece.org./trade/timber/tim-fact/rus.htm

Uusivuori, J. and Buongiorno, J. (1991) Pass-through of exchange rates on prices of forest product exports from the United States to Europe and Japan. *Forest Science* 37, no. 3, 931-948.

Vorob'yev, G. I., Moiseyev, N. A., Lositsskii, S. G., Tsekhmistrenko, A. F., Feofilov, V. A., Tsimek, A. A., Pobyedinskii, A. V. and Yurkevich, M. S. (1979) *Ekonomicheskaya geografiya lesnikh resursov SSSR*. Lesnaya Promishlennost', Moscow.

Wakimoto, Y. (1998) Juutakukenchiku wa choukiteimeijidai ni totsunyuu ka. *Mokuzai Jouhou*, no. 8, 1-9.

Weidenbeck, J. K. and Araman, P. A. (1993) Possible demands for eastern hardwoods resulting from harvest restrictions in the Pacific Northwest. *Forest Products Journal* 43, no. 10, 51-57.

White, A. (1997) Shinrin kara mokuzai kakousangyou e. *Mokuzai Jouhou*, no. 6, 24-27.

Yagodnikov, Yu. A. and Mikhailov, G. M. (1991) *Leso-zagotovitel'noe proizvodstvo*. Lesnaya Promishlennost', Moscow.

Yakunin, A. G., Shlikov, V. M. and Grachov, V. N. (1987) *Lesnaya industriya dal'nevo vostoka*. Lesnaya Promishlennost', Moscow.

Yanasachi, H. (1997) Jinkourinzai yushutsukoku to shite no Nyuu Jirando to Chiri. *Mokuzai Jouhou*, no. 12, 1-17.

Yorimitsu, R. (1992) Thriving housing construction in Japan and the factors contributing to it. In: Handa, R. *et al.* (eds) *The Current State of Japanese Forestry VIII*. The Japanese Forest Economics Society, Tokyo, pp. 1-16.

Yoshida, M. (1981) Repercussion effects of economic policy on demand for lumber and plywood. In: Handa, R. *et al.* (eds.) *The Current State of Japanese Forestry I*. The Japanese Forest Economics Society, Tokyo, pp. 1-6.

Zenkoku Shinrin Kumiai Rengoukai (1991) *Shinrinkumiai katsudou no genjou to kadai*. Zenkoku Shinrin Kumiai Rengoukai, Tokyo.

Zhang, Y., Buongiorno, J. and Zhu, S. (1997) *Trends and outlook for forest products consumption, production and trade in the Asia Pacific region*. Asia Pacific Forest Outlook Study Working paper APFSOS/WP/12, FAO, Rome.

Index